Bayesian Compendium

Marcel van Oijen

Bayesian Compendium

 Springer

Marcel van Oijen
Edinburgh, UK

ISBN 978-3-030-55899-4   ISBN 978-3-030-55897-0   (eBook)
https://doi.org/10.1007/978-3-030-55897-0

This Springer imprint is published by the registered company Springer Nature Switzerland AG
The registered company address is: Gewerbestrasse 11, 6330 Cham, Switzerland

Preface

Why This Book?

The recurring topic during my 35 years in science has been a struggle with uncertainties. My primary field is agricultural, environmental and ecological science using both process-based and statistical models. And in each of my studies, uncertainties in data and models have kept cropping up. I found that the literature contains many ways for classifying and analysing these uncertainties, but they mostly seem based on arbitrary criteria: a confusing menagerie of methods.

My encounters with probability theory, and with Bayesian methods in particular, showed that there is no need for confusion; there is a general perspective that can be applied in every case. The online papers by Edwin Jaynes, and his posthumously published book (2003), followed by the more practice-oriented small book by Sivia (2006) did much to clarify matters. Very striking was Jaynes' explanation of the Cox postulates, which proved that consistent rational thinking requires the use of the rules of probability theory. Jaynes and Sivia took many of their examples from physics, but it was clear that Bayesian probabilistic methods are completely generic and can be applied in any field. It also became clear that the methods do come with some practical problems. They require researchers to think carefully about their prior knowledge before embarking on an analysis. And the Bayesian analysis itself tends to be computationally demanding. Much research effort has gone into resolving these issues, and good methods are available.

So I have happily employed Bayesian methods during the past two decades. However, when teaching the approach to others, I found that there was no short yet comprehensive introductory text for scientists who do not have much statistical background. Especially amongst dynamic modellers, the study of statistical methods is not given high priority. But this book is intended to provide a quick and easy entry into Bayesian methods to all modellers, irrespective of the types of models and data that they work with.

Who is This Book for?

This book is primarily for scientists who are newcomers to Bayesian methods, and who want to use data to parameterise and compare their models while accounting for uncertainties in data, model parameters and model structure. There are of course excellent books on Bayesian methods for beginners, but they tend to be very long texts that neglect the dynamic modeller. There is a need for short and easily digestible descriptions of Bayesian methods that cover the wide range of process-based and empirical models that are in common use. All chapters in this book are short, so it is more a compendium than a detailed manual.

The book is for every scientist who works with models and data, irrespective of the type of model or data. Although aimed at beginners, I hope that more advanced users of Bayesian methods will also find material of interest in this book. They may be unfamiliar with some of the more specialized topics, or find the book useful for teaching. They may also be interested in the way this book tries to clarify the terminological confusion that reigns in scientific modelling and in applied statistics.

What is in the Book?

The book begins with a gentle introduction to Bayesian thinking. Once the basic ideas are in place, it provides explanations of a very wide range of modelling methods in three languages: English, mathematics and R. I believe that a mixture of the three languages, complemented with figures, is the best way to teach quantitative science. You may find English too woolly, mathematics too compact or R-code too problem-specific. But together they may provide clear, reasonably short and practical explanations. Some exercises are provided as well.

The code examples in this book are all short, and they mostly use basic R that can easily be understood even if you are used to programming in a different computer language. R does come with a wealth of optional add-ons, called R-packages, but those will be used only occasionally. All essential code is in the book itself, and the complete code is available online (https://www.springer.com/de/book/9783030558963).

Almost every example will make use of very simple artificial data sets. I mostly leave out the messiness of real data sets because they are distracting and obscure the view of how the methods work. Likewise, I focus on brevity and simplicity of code, rather than on speed of execution. The aim is to explain the diversity of Bayesian computational techniques that you can find in the literature. The algorithms that we show, such as MCMC, can be used with any kind of model and data. In fact, one of the main advantages of Bayesian statistical modelling is that all idiosyncrasies of data, such as missing data, can be easily handled.

Outline of Chapters

This book has six sections, with the following contents:

- *The basics*: Chaps. 1–12.

  - In these chapters, we introduce the basics of Bayesian thinking to you. In terms of abstract concepts, there is not too much to learn: it's all about using a prior distribution and likelihood function to derive the posterior distribution. For simple models, the mathematics is very easy. But practical application of Bayesian methods can be harder if our models or data are complicated. We then rely on computer algorithms to find representative samples from the posterior distribution.

- *Advanced material*: Chaps. 13–18.

  - These chapters introduce methods that help us with more complicated prior distributions and model structures. We also show how Bayesian thinking helps with risk analysis and decision making.

- *Specific models*: Chaps. 19–23.

  - The models that we discuss in these chapters are quite simple functions, or built from combinations of simple functions. But the models are highly versatile and widely applicable, even in situations where we do not know beforehand what the shape of our function of interest is!

- *Outlook to the future*: Chap. 24.

  - This chapter reviews the state of Bayesian modelling and identifies trends of future methodological development. We give a few pointers for further self-study.

- *Appendices with tips and tricks.*

  - In these appendices, we present the notation we use in this book, and very briefly summarize relevant mathematics, probability theory and software.

- *Solutions to exercises, index, references.*

Acknowledgements

Science is a highly collaborative affair, and over the years I have discussed Bayesian methods with a great number of people. I believe that most of them know how grateful I am to them, so I will do some cluster analysis and thank people by group—mentioning just a few names explicitly. I thank many statisticians in the Netherlands and the UK, who taught me over the years, including Ron Smith and Jonathan Rougier who were co-authors on my first Bayesian paper. To the present day, my publications with a Bayesian or otherwise probabilistic focus have been with about 140 different co-authors: a large group of inspiring people! Further inspiration came from colleagues in projects Nitro-Europe, Carbo-Extreme, GREENHOUSE, MACSUR, UK-SCaPE, PRAFOR and various COST-Actions, and from the Cambridge gang of July 2019. I thank my colleagues in the UK, Netherlands, Belgium, Germany, France, Portugal, Hungary, Norway and Finland who invited me to teach, lecture or work on Bayesian methods, and the many students. Mats Höglind gave me the opportunity for a sabbatical year in Norway where I spent much time with Edwin Jaynes. I enjoyed organizing a Bayesian course in Edinburgh because of the happy collaboration with Lindsay Banin, Kate Searle, David Cameron and Peter Levy, and the enthusiasm of the students. David and Peter have been my main discussion partners over the years, so my extra special thanks go to them.

Finally, with more than gratitude, my wife Netty.

Edinburgh, UK   Marcel van Oijen

Contents

1 Introduction to Bayesian Thinking 1
  1.1 Bayesian Thinking 1
  1.2 A Murder Mystery 2
  1.3 Bayes' Theorem 3
    1.3.1 Implications of Bayes' Theorem 4
    1.3.2 The Odds-Form of Bayes' Theorem and a Simple
        Application 5

2 Introduction to Bayesian Science 7
  2.1 Measuring, Modelling and Science: The Three Basic
    Equations 7
  2.2 Terminological Confusion 8
  2.3 Process Based Models Versus Empirical Models 11
  2.4 Errors and Uncertainties in Modelling 12
    2.4.1 Errors and Uncertainties in Model Drivers 12
    2.4.2 Errors and Uncertainties in Model Parameters 13
    2.4.3 Errors and Uncertainties in Model Structure 13
    2.4.4 Forward Propagation of Uncertainty to Model
        Outputs 14
  2.5 Bayes and Science 15
  2.6 Bayesian Parameter Estimation 15

3 Assigning a Prior Distribution 17
  3.1 Quantifying Uncertainty and MaxEnt 18
  3.2 Final Remarks for Priors 20

4 Assigning a Likelihood Function 23
  4.1 Expressing Knowledge About Data Error in the Likelihood
    Function 25
  4.2 What to Measure 27

5   Deriving the Posterior Distribution 29
   5.1   Analytically Solving Bayes' Theorem: Conjugacy 29
   5.2   Numerically 'Solving' Bayes' Theorem: Sampling-Based
         Methods 31
   Exercises 32

6   Sampling from Any Distribution by MCMC 33
   6.1   MCMC 33
   6.2   MCMC in Two Lines of R-Code 34
   6.3   The Metropolis Algorithm 35
   Exercise 38

7   Sampling from the Posterior Distribution by MCMC 39
   7.1   MCMC and Bayes 39
         7.1.1   MCMC and Models 39
         7.1.2   The Need for Log-Transformations in MCMC 40
   7.2   Bayesian Calibration of a 2-Parameter Model Using
         the Metropolis Algorithm 41
         7.2.1   The Metropolis Algorithm 41
         7.2.2   Failed Application of MCMC Using the Default
              Settings 42
   7.3   Bayesian Calibration of a 3-Parameter Model Using
         the Metropolis Algorithm 44
   7.4   More MCMC Diagnostics 44
   Exercises 46

8   Twelve Ways to Fit a Straight Line 49
   8.1   Hidden Equivalences 49
   8.2   Our Data 50
   8.3   The Normal Equations for Ordinary Least Squares
         Regression (OLS) 50
         8.3.1   Uncertainty Quantification 51
   8.4   Regression Using Generalised Least Squares (GLS) 52
         8.4.1   From GLS to WLS and OLS 53
   8.5   The Lindley and Smith (LS72) Equations 53
   8.6   Regression Using the Kalman Filter 54
   8.7   Regression Using the Conditional Multivariate Gaussian 55
   8.8   Regression Using Graphical Modelling (GM) 56
   8.9   Regression Using a Gaussian Process (GP) 57
   8.10 Regression Using Accept-Reject Sampling 58
   8.11 Regression Using MCMC with the Metropolis Algorithm 58
   8.12 Regression Using MCMC with Gibbs Sampling 58
   8.13 Regression Using JAGS 60
   8.14 Comparison of Methods 61
   Exercises 61

9 MCMC and Complex Models 63
   9.1 Process-Based Models (PBMs) 63
       9.1.1 A Simple PBM for Vegetation Growth: The
              Expolinear Model 63
   9.2 Bayesian Calibration of the Expolinear Model 65
   9.3 More Complex Models 66

**10 Bayesian Calibration and MCMC: Frequently Asked
    Questions** 69
   10.1 The MCMC Algorithm 69
   10.2 Data and Likelihood Function 70
   10.3 Parameters and Prior 72
   10.4 Code Efficiency and Computational Issues 72
   10.5 Results from the Bayesian Calibration 74

11 After the Calibration: Interpretation, Reporting, Visualization . . . 77
   11.1 Interpreting the Posterior Distribution and Model
        Diagnostics 77
   11.2 Reporting 79
   11.3 Visualising Uncertainty 80

12 Model Ensembles: BMC and BMA 81
   12.1 Model Ensembles, Integrated Likelihoods and Bayes
        Factors 81
   12.2 Bayesian Model Comparison (BMC) 82
   12.3 Bayesian Model Averaging (BMA) 83
   12.4 BMC and BMA of Two Process Based Models 84
       12.4.1 EXPOL5 and EXPOL6 84
       12.4.2 Bayesian Calibration of EXPOL6's Parameters 85
       12.4.3 BMC and BMA of EXPOL5 and EXPOL6 85

13 Discrepancy 89
   13.1 Treatment of Discrepancy in Single-Model Calibration 90
   13.2 Treatment of Discrepancy in Model Ensembles 91

14 Gaussian Processes and Model Emulation 93
   14.1 Model Emulation 93
   14.2 Gaussian Processes (GP) 94
   14.3 An Example of Emulating a One-Input, One-Output Model . . . 96
       14.3.1 Analytical Formulas for GP-calibration
              and Prediction 96
       14.3.2 Using R-Package geoR for GP-calibration
              and Prediction 98
   14.4 An Example of Emulating a Process-Based Model
        (EXPOL6) 100

14.4.1 Training Set 100
14.4.2 Calibration of the Emulator 101
14.4.3 Testing the Emulator 102
14.5 Comments on Emulation 104
Exercise 105

15 Graphical Modelling (GM) 107
15.1 Gaussian Bayesian Networks (GBN) 107
15.1.1 Conditional Independence 108
15.2 Three Mathematically Equivalent Specifications
of a Multivariate Gaussian 109
15.2.1 Switching Between the Three Different
Specifications of the Multivariate Gaussian 110
15.3 The Simplest DAG Is the Causal One! 112
15.4 Sampling from a GBN and Bayesian Updating 112
15.4.1 Updating a GBN When Information About Nodes
Becomes Available 112
15.5 Example I: A 4-Node GBN Demonstrating DAG Design,
Sampling and Updating 114
15.6 Example II: A 5-Node GBN in the form of a Linear Chain 116
15.7 Examples III & IV: All Relationships in a GBN are Linear 116
15.7.1 Example III: A GBN Representing Univariate
Linear Dependency 117
15.7.2 Example IV: A GBN Representing Multivariate
Stochastic Linear Relations 117
15.8 Example V: GBNs can do Geostatistical Interpolation 118
15.9 Comments on Graphical Modelling 119
Exercises 120

16 Bayesian Hierarchical Modelling (BHM) 121
16.1 Why Hierarchical Modelling? 122
16.2 Comparing Non-hierarchical and Hierarchical Models 123
16.2.1 Model A: Global Intercept and Slope, Not
Hierarchical 125
16.2.2 Model B: Cv-Specific Intercepts and Slopes,
Not Hierarchical 126
16.2.3 Model C: Cv-Specific Intercepts and Slopes,
Hierarchical 126
16.2.4 Comparing Models A, B and C 127
16.3 Applicability of BHM 128
Exercise 128

17 Probabilistic Risk Analysis and Bayesian Decision Theory 129
17.1 Risk, Hazard and Vulnerability 129
17.1.1 Theory for Probabilistic Risk Analysis (PRA) 130

17.2   Bayesian Decision Theory (BDT) 131
       17.2.1   Value of Information 132
17.3   Graphical Modelling as a Tool to Support BDT 133

18   Approximations to Bayes 135
18.1   Approximations to Bayesian Calibration.................. 136
18.2   Approximations to Bayesian Model Comparison............ 136

19   Linear Modelling: LM, GLM, GAM and Mixed Models 137
19.1   Linear Models 137
19.2   LM 138
19.3   GLM 138
19.4   GAM 139
19.5   Mixed Models 139
19.6   Parameter Estimation 139
       19.6.1   Software 140

20   Machine Learning.. 141
20.1   The Family Tree of Machine Learning Approaches......... 142
20.2   Neural Networks..................................... 143
       20.2.1   Bayesian Calibration of a Neural Network 145
       20.2.2   Preventing Overfitting......................... 147
20.3   Outlook for Machine Learning......................... 148
Exercises 149

21   Time Series and Data Assimilation 151
21.1   Sampling from a Gaussian Process (GP) 151
21.2   Data Assimilation Using the Kalman Filter (KF) 154
       21.2.1   A More General Formulation of KF 157
21.3   Time Series, KF and Complex Dynamic Models 159
Exercises 160

22   Spatial Modelling and Scaling Error........................ 161
22.1   Spatial Models 161
22.2   Geostatistics Using a GP............................. 162
22.3   Geostatistics Using geoR 163
22.4   Adding a Nugget 164
22.5   Estimating All GP-hyperparameters 165
22.6   Spatial Upscaling Error 166

23   Spatio-Temporal Modelling and Adaptive Sampling............ 169
23.1   Spatio-Temporal Modelling 169
23.2   Adaptive Sampling 170
23.3   Comments on Spatio-Temporal Modelling and Adaptive
       Sampling 172

24   What Next?   173
     24.1   Some Crystal Ball Gazing   174
     24.2   Further Reading   175
     24.3   Closing Words   176

Appendix A: Notation and Abbreviations   177

Appendix B: Mathematics for Modellers   179

Appendix C: Probability Theory for Modellers   181

Appendix D: R   185

Appendix E: Bayesian Software   187

Appendix F: Solutions to Exercises   189

References   193

Index   201

Chapter 1
Introduction to Bayesian Thinking

In recent years, there has been a trend toward basing scientific research under conditions of incomplete information, i.e. most of science, on probability theory (e.g. Hartig et al. 2012; Jaynes 2003; Ogle and Barber 2008; Sivia and Skilling 2006; Van Oijen et al. 2011). This is the approach that we take in this book too. We aim to show how defining all uncertainties in modelling as probability distributions allows for rigorous reduction of those uncertainties when new data become available. The approach that we are presenting is known in the literature under many different names, including *Bayesian calibration*, *data assimilation*, *model-data fusion* and *inverse modelling*. Whilst the different names refer to different applications of modelling, they all share the idea of specifying probability distributions which are modified according to the rules of probability theory (in particular, *Bayes' Theorem*) when new data come in. It is this idea that facilitates the comprehensive analysis of errors and uncertainties.

Lindley (1991) stated the importance of probability theory as follows: "Probability, it has been said, is merely common sense reduced to calculation. It is the basic tool for appreciating uncertainty, and uncertainty cannot be adequately handled without a knowledge of probability." And it is possible to show formally that rational, coherent thinking implies using the rules of probability theory Jaynes (2003).

1.1 Bayesian Thinking

The basics of Bayesian thinking are simple. There are just three elements, connected by probability theory. The elements are: (1) your prior belief about a quantity or proposition, (2) new information, (3) your posterior belief. Probability theory provides the logical connection from the first two elements to the last. So all we need to learn is how to express beliefs and new information in the form of probability

© Springer Nature Switzerland AG 2020
M. van Oijen, *Bayesian Compendium*,
https://doi.org/10.1007/978-3-030-55897-0_1

distributions, and then we can simply follow the rules of probability theory. That is all!

While the simplicity of Bayesian thinking is a given, that does not mean it is necessarily easy to learn. Consistently thinking in terms of probability theory is not second nature to everyone. And there is no unique simplest way to teach Bayesian thinking to you. It all depends on your background, and your favourite mode of rational thinking. Do you prefer to begin with abstract concepts, then mathematical equations, then examples? Or do you wish to begin with puzzles or anecdotes, and learn how they can all be approached in the same way? Perhaps you like to start from your knowledge of classical statistics and learn how its methods can always be interpreted, and often improved, in a Bayesian way? But here we begin with a short detective story ...

1.2   A Murder Mystery

You are called to a country house: the owner has been found murdered in the library. The three possible suspects are his wife, his son, and the butler. *Before reading on, who do you believe committed the crime?* And do not say: "I can't answer that, I have not inspected the evidence yet." You are a Bayesian detective, so you can state your prior probabilities.

Your assistant says "I bet it's the wife", but you find that silly bias. You see the butler as the prime suspect, and would give odds of 4:1 that he is the culprit. You find the wife just as improbable as the son. So your prior probability distribution for butler-wife-son is 80–10–10%. Of course, you would not really bet money on the outcome of the case—you're a professional—but you decide to investigate the butler first. To your surprise, you find that the butler has a perfect alibi. *What is your probability distribution now?* The answer is that the alibi of the butler has no bearing on the wife or son, so they remain equally likely candidates, and your probability distribution becomes 0–50–50%.

Next you inspect the library, and find that the murder was committed with a blunt instrument. *How does that change your probabilities?* You assess the likelihood of such a murder weapon being chosen by a man to be twice as high as by a woman. So that changes your probabilities to 0–33–67%.

I leave you to finish the story to a logical conclusion where sufficient evidence has been processed such that the murderer is identified beyond reasonable doubt. But what does this murder mystery teach us? Well, our story is fiction, but it contains the three steps of any Bayesian analysis: assigning a prior probability distribution, acquiring new information, updating your probability distribution following the rules of probability theory. Everything in this book, and in Bayesian statistics generally, is about one or more of these three steps. So if you found the detective's reasoning plausible, then you are already a Bayesian! In fact, there is a rich literature in the field of psychology that shows that human beings at least to some extent make decisions in a Bayesian way.

Exercise 1.1 You saw that your assistant started with a different prior distribution for the murderer. Assuming he interprets the new information about the murder weapon in the same way as you do, what is his posterior distribution? If you would keep finding new evidence, would your assistant and you eventually end up with the same 'final' posterior distribution?

1.3 Bayes' Theorem

Bayes' Theorem dates back to the 18th century (Bayes 1763), and its use in science has been explained in textbooks (e.g. Gelman et al. 2013; McElreath 2016) and tutorial papers (e.g. Van Oijen et al. 2005; Hartig et al. 2012). So what is the theorem? To state it, we need to introduce some notation (you can find more in the Appendices). We write $p[A|B]$ for the *conditional probability* of A being true given that B is true. The theorem is then written as follows:

BAYES' THEOREM:

$$p[A|B] = \frac{p[A]\, p[B|A]}{p[B]}. \tag{1.1}$$

All four terms in Eq. (1.1) are probabilities or probability densities, and the theorem is true irrespective of what is denoted by A and B. A is what we are interested in. It can stand for "it will rain tomorrow" or "crop yield will exceed 10 tons per hectare" or "the biomass density parameter has value 900 (kg m^{-3}). We want to quantify the probability that the statement A is true given some information B, and we write this conditional probability as $p[A|B]$. In scientific applications, A often refers to a parameter or parameter vector whose true value θ we want to know, so A could stand for "$\theta = 1$", or "$\theta = 2$", or any other parameter value. And B could then be the statement "my sensor gives me the value y". Note that in such cases, Bayes' Theorem actually gives us infinitely many different correct probability statements at once, because it remains valid for every possible parameter value θ and any data y. In other words, Bayes' Theorem defines the whole conditional probability *distribution* for the parameter value given the data. Here is Bayes' Theorem again using these more common symbols:

BAYES' THEOREM (common notation):

$$p[\theta|y] = \frac{p[\theta]p[y|\theta]}{p[y]}. \tag{1.2}$$

All four terms in the theorem have a name. $p[\theta|y]$ is the *posterior probability distribution* for the parameter θ, $p[\theta]$ is its *prior distribution*, $p[y|\theta]$ is the *likelihood function*, and $p[y]$ is the *evidence*. Your job in a Bayesian analysis will be to define the prior and likelihood, and the evidence will then automatically be defined as well (as we'll explain later), after which Bayes' Theorem gives us the posterior distribution that we are seeking. How to define the prior and likelihood will be explained in Chaps. 3 and 4.

Exercise 1.2 We just gave names to the four terms in Bayes' Theorem. We referred to $p[\theta|y]$ and $p[\theta]$ as 'probability distributions'. That is, strictly speaking, incorrect. Why?

We can also write Bayes' Theorem with just three terms. In most applications, the value of the data y is given, and we cannot change that. But we can consider many different possible values of the parameter θ. So $p[y]$ is constant but the other three terms vary depending on the value of θ. In terms of the varying quantities, we can thus say that the posterior is proportional to the product of the prior and the likelihood:

$$\text{BAYES' THEOREM (proportionality form):}$$
$$p[\theta|y] \propto p[\theta]p[y|\theta]. \tag{1.3}$$

1.3.1 Implications of Bayes' Theorem

Note that in our story, we allowed for the possibility that you, the detective, and your assistant assign different prior probabilities of guilt to the three suspects. This subjective aspect of probability in fact applies to all prior probabilities and likelihood functions that we assign informally. Different people have different expertise and background information, so we can expect them to assign different probabilities to events in the real world. But once we have specified our prior and likelihood, we have no choice but to accept the posterior distribution that Bayes' Theorem then gives us: there is no subjectivity in that step. In a scientific context, we need to account for all inevitable subjectivities by being transparent about them (i.e. mention which prior and likelihood we used), and where possible minimise the subjectivity by using formal methods of specifying the prior distribution. We shall come back to that issue in Chap. 3.

Bayes' Theorem as written in (1.2) or (1.3) has some immediate consequences. First, if you assign a prior probability of zero to certain values of the parameters θ, then whatever evidence may subsequently appear, those parameter values will always remain impossible. The product of zero and any other value is always zero. For example, if you have parameters that a priori are considered to be positive, then even a plethora of negative measurement values will not lead to negative estimates. Likewise, once you have found evidence that implies zero likelihood for certain parameter values, then those values remain excluded for ever. So we should only rule out parameter values—in their prior or likelihood function—if we are 100% certain that those values are impossible. On the other hand, if we do have that certainty, then a restricted prior can make subsequent analysis much more efficient.

We now have the formal apparatus to tackle problems in a Bayesian way. Let's briefly return to our earlier detective story, and see how Bayes' Theorem fits in. The story is a problem of parameter estimation. There is only one parameter, θ, with just three possible values: $\theta \in \{wife, son, butler\}$. So our prior is a discrete probability distribution. The first piece of evidence is the alibi of the butler, which equates—in probabilistic language - to a likelihood of zero for $\theta = butler$. If the butler were guilty, then that evidence could not have been found. We write that symbolically as $p[alibi|\theta = butler] = 0$. You can now rewrite the story yourself as a sequence of applications of Bayes' Theorem where the posterior distribution after the first piece of evidence becomes the prior when processing the second piece of information, and so on. Bayes formalizes the process of learning.

1.3.2   The Odds-Form of Bayes' Theorem and a Simple Application

At this point, we show yet another mathematically equivalent form of Bayes' Theorem which is especially useful when our parameter can have only two possible values, for example 0 or 1, true or false, wife or son. In such cases, the odds-form of Bayes' Theorem is useful. The odds on an event are the probability of it happening divided by the probability of it not happening: $O[event] = p[event]/(1 - p[event])$. Let's derive the posterior odds for a parameter $\theta \in \{a, b\}$. We use shorthand notation: instead of writing $p[\theta = a]$, we write $p[a]$ ctc. So the prior odds on a are $O[a] = p[a]/(1 - p[a]) = p[a]/p[b]$. Then:

BAYES' THEOREM (odds form):

$$p[a|y] = \frac{p[a]\,p[y|a]}{p[y]}$$

$$p[b|y] = \frac{p[b]\,p[y|b]}{p[y]} \tag{1.4}$$

$$\implies \frac{p[a|y]}{p[b|y]} = \frac{p[a]}{p[b]}\frac{p[y|a]}{p[y|b]}$$

$$\implies O[a|y] = O[a]\frac{p[y|a]}{p[y|b]}.$$

So this gives us a very simple way of expressing Bayes' Theorem: *posterior odds are prior odds times the likelihood ratio*. Let's use this odds-form of Bayes' Theorem in a simple application. We have already used Bayesian thinking in our murder mystery, but now we turn to an example where Bayes' Theorem is used more formally and quantitatively. It is the perhaps most popular example used in Bayesian tutorials: how to interpret a medical diagnosis. Say that an epidemic has started and 1% of people is infected. A diagnostic test exists that gives the correct result in 99% of cases. We have the test done, and the result is positive, which is bad news because for diagnosticians 'positive' means 'infected'. But what is the probability that you actually do have the disease?

Exercise 1.3 Before reading on, answer the question using the odds-form of Bayes' Theorem.

We now give the answer. As for every Bayesian analysis we must specify our prior and likelihood. Our prior probability of being infected is 1% so our prior odds are $O[infected] = 1/99$. Our likelihood function is: $\{p[positive|infected] = 99\%$, $p[positive|not\ infected] = 1\%\}$, so our likelihood ratio is 99. Your posterior odds for having the disease are thus $O[infected|positive] = 1/99$ times $99 = 1$. These even odds mean that, perhaps surprisingly giving the high accuracy of the diagnostic test, your posterior probability of having the disease is still only 50%. This is because the small error probability of the test is matched by an equally small prior probability of having the disease.

Exercise 1.4 Could we also use an odds-formulation of Bayes' Theorem in case the parameter has more than two possible values? For example, in our murder mystery, our prior odds for {butler:wife:son} were {80%:10%:10%} = {8:1:1}. Would it be correct to say that new information y with likelihood ratios $\{p[y|butler] : p[y|wife]$ $: p[y|son]\} = \{b{:}w{:}s\}$ would have given us posterior odds of $\{8b{:}w{:}s\}$?

Chapter 2
Introduction to Bayesian Science

2.1 Measuring, Modelling and Science: The Three Basic Equations

In science, we use models to help us learn from data. But we always work with incomplete theory and measurements that contain errors. The consequence of this is that we need to acknowledge uncertainties when trying to extract information from measurements using models. Not a single measurement, not a single explanation or prediction is wholly correct, and we can never know exactly how wrong they are. This book is about how to best extract information from data using models, acknowledging all relevant uncertainties. Our overall approach is captured in three equations. The first is the *Measurement Equation*, which acknowledges observational error:

MEASUREMENT EQUATION:
$$z = y + \epsilon_y,$$

$$(2.1)$$

where z is the true value, y is the measured value, and ϵ_y is measurement error. The second equation is the *Modelling Equation*, and it expresses the presence of a different kind of error:

MODELLING EQUATION:
$$z = f(x, \theta) + \epsilon_{model},$$

$$(2.2)$$

where f is the model, x are the environmental conditions for which we run the model, θ are the model's parameter values, and the final term ϵ_{model} represents model error which arises because f, x and θ will all be wrong to some extent. All the terms in the above equations can be scalars (single values) but more commonly they are vectors (multi-dimensional). Depending on the discipline, and sometimes the individual author or problem, the environmental conditions x are variously referred to as drivers, covariates, predictors, boundary conditions, explanatory or independent

© Springer Nature Switzerland AG 2020
M. van Oijen, *Bayesian Compendium*,
https://doi.org/10.1007/978-3-030-55897-0_2

variables. Their defining characteristic is that we treat them as given rather than trying to estimate them.

When we combine the Measurement and Modelling equation we arrive at the *Science Equation*:

$$\underline{\text{SCIENCE EQUATION:}}$$
$$y = f(x, \theta) + \epsilon_{model} + \epsilon_y. \tag{2.3}$$

Science is the attempt to explain observations y using a model f where we try to minimize model error ϵ_{model} and measurement error ϵ_y so that the model can be used to make predictions about the real world (z). The Bayesian approach to science recognizes and quantifies uncertainties about all six elements of the Science Equation: $y, f, x, \theta, \epsilon_{model}$, and ϵ_y. We shall return to these equations at various points in the book (and expand them), but the main thing to notice for now is that we recognize errors in both measurement and modelling, which makes it hard to find the best parameter values θ and the best model form f.

2.2 Terminological Confusion

Because this book focuses on uncertainties in measuring and modelling, we need to adopt language from probability theory and statistics. The problem is that many concepts have been given very confusing names. Perhaps the most confusing term for scientists is *random variable*. Take the commonly heard statement "To Bayesians, every parameter is a random variable that obeys a specific probability distribution." That statement is not strictly wrong but it is highly misleading. It suggests that parameters can jump around, assume arbitrary values. But Bayesians do not believe that. They believe that all parameters have specific values that we, however, can never know with infinite precision. So a random variable is just a real-world property about which we have incomplete knowledge. Randomness says nothing about the property, it says something about the state of knowledge of the researcher. And that state of knowledge will differ between people, so different researchers will quantify their uncertainties with different probability distributions. We shall try to avoid the term in this book and only talk about uncertainties.

Of course, terminology can never be perfect. This book combines work from mathematics, logic, probability theory & statistics, computer science & programming, and the sciences. These disciplines often use the same terms with different meanings, and they often use different terms to refer to the same thing. Even the most basic terms such as *model*, *parameter* and *process* have different definitions in the different fields. The following list defines our use of these and other terms.

- *Error, residual, uncertainty, variability*. These are terms that have different meanings, but are very often confused.

– *Error* is defined as the difference between an estimate (measured or modeled) and the true value. So in Eq. (2.1), the term ϵ_y represents measurement error, and in Eq. (2.2), ϵ_{model} represents modelling error.

– A *residual* is the difference between a measured and a modelled value. Referring to the same two equations, the residuals are defined as $y - f(x, \theta)$.

– *Uncertainty* is having incomplete knowledge about a quantity. Uncertainty can always be represented by a probability distribution, denoted as $p[..]$. So the probability distribution $p[\epsilon_y]$ defines which values of measurement error we deem possible and their relative probabilities of occurring. Say our instrument is accurate (unbiased) but has only 50% precision. Then we are highly uncertain about the error and might write $p[\epsilon_y] = N[\epsilon_y; \mu = 0, \sigma = y/2]$, which stands for a normal (or 'Gaussian') distribution with mean zero and standard deviation equal to half the observed value. Conditional probability distributions, e.g. for θ given a certain value of y, are denoted in the standard way as $p[\theta|y]$. Note that we use square brackets for probabilities, probability distributions and likelihoods (e.g. $p[..]$, $N[..]$) and parentheses for functions such as $f(..)$.

– *Variability* is the presence of differences in a set of values. Variability can lead to sampling error of uncertain magnitude. Say we randomly sample 100 trees from a large forest, measure stem diameter on each, and collect the 100 values in vector y. The standard deviation σ_y is then a measure of the variability in the sample. If all measurements are very precise then there will be little uncertainty about that variability. But if we use the sample mean as an estimate for the mean diameter of all trees in the forest, then there will be an unknown sampling error. In this example, the uncertainty about the sampling error may well be represented by a normal distribution with mean zero and standard deviation equal to $\sigma_y / \sqrt{100}$.

• *Model, input, output.* Every *model* is a function that operates on inputs and converts them into outputs. Typical inputs are *parameters* and independent variables. Inputs as well as outputs can be scalars (one-dimensional) but are more often vectors (multi-dimensional). Independent variables can be time, spatial coordinates or environmental drivers such as time series of weather variables or human interventions to the modelled system. Models are used for two purposes: to explain the past and to predict the future. In both cases, the model is just a tool to extract information from measurements. Analysis of variance, for example, is a statistical model for extracting information from multi-factorial experiments. A complex global climate model is a tool to extrapolate past measurements of weather and other parts of the earth system to future climate change. There is no fundamental difference between the ANOVA and the GCM, both are imperfect data processors that supposedly target causal relationships in nature. The only difference is that the GCM is more complex.

– *Dynamic model.* A model with time as a variable.
– *Process-based model (PBM).* PBMs are deterministic dynamic models that simulate real-world processes. They typically have two types of input: drivers and

parameters. Drivers are environmental variables (weather, disturbances, management) whereas parameters are constant properties of the system.

– *Empirical model*. A model that does not dynamically simulate processes. So in this book, we use PBMs and empirical models as complementary terms. Every model belongs to either the first or the second category.
– *Deterministic model*. A model that for any specific input always produces the same output.
– *Stochastic model*. A model that produces different output every time it is run, even if the input remains the same. It is a complementary term to deterministic model: every model is either the one or the other.
– *Statistical model*. A model that explicitly represents uncertainty about errors by means of probability distributions. Both PBMs and empirical models can be embedded in a statistical model. This book shows how a statistical model can be wrapped around any model, irrespective of whether the core-model is process-based or empirical, deterministic or stochastic (Fig. 2.1).

- *Parameter*. Here treated as 'constant', not as 'measurable property'. A parameter is a constant input to a model. Typical examples of parameters are rate-values of processes and the initial values of state variables.
- *Prior, likelihood, posterior*. See Chap. 1.
- *Probability, random*. We already stated that randomness is in the eye of the beholder; believing it to be a property of the world is, in the words of Edwin Jaynes (Jaynes 2003), a "mind-projection fallacy". Likewise, probability is not a frequency but a state of mind. But it is fine to say that we (not nature!) sample a random value from a probability distribution (although even there we are ignoring the fact that a computer can only produce pseudo-random numbers). For example, we may want to sample from the prior distribution for the parameters of a model to see what our prior parameter uncertainty implies for our prior predictive uncertainty.
- *Process*. In most cases this will refer to a physical process in the real world. But there is a very different statistical meaning to this term as well, in the context of 'stochastic process', which is a probability distribution for the shape of a function. *Gaussian Processes* are an example of the latter, and they will feature quite prominently in this book (e.g. Chaps. 14 and 21). Both meanings of the term 'process' are so engrained in their fields that we will have to keep using both.

For now, these terms suffice, but we shall gradually be introducing more terminology.

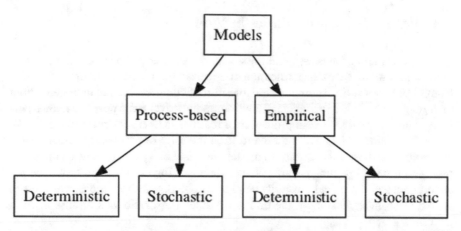

Fig. 2.1 The four model types. Each of them can be embedded in a statistical model. Each of them should be

2.3 Process-Based Models Versus Empirical Models

All methodology that we present in this book can be applied to any kind of model: simple or complex, site-based or area-based, deterministic or stochastic, process-based or empirical. Most books that explain Bayesian methods use examples of small data sets and fairly simple models, and we shall in most cases do so too. It is a valid approach because the Bayesian thinking (define prior & likelihood, derive posterior) is the same in all cases. However, the complexity of a model does affect the ease of Bayesian computation. And therefore, some of our examples shall be drawn from the literature on deterministic process-based modelling of ecosystems. This is a class of dynamic models that simulates vegetation growth as part of the carbon-, water- and nutrient-cycles in the soil-tree-atmosphere system (Reyer 2015). Such process-based models (PBMs) can be used for assessment of multiple ecosystem services (Mäkelä et al. 2012). PBMs tend to be parameter-rich and computationally demanding. We shall explain how these characteristics hamper the comprehensive application of the probabilistic approach to uncertainties, and which solutions have been proposed. PBMs are particularly useful for the study of complex adaptive systems of which important internal processes and mechanisms are known, so we do not need to treat them as black boxes. PBMs aim to represent real mechanisms, i.e. causal relationships. You can say that empirical models do so too but only half-heartedly.

The literature offers the modeler a bewildering variety of techniques and approaches, with grand names such as Data Assimilation, Model-Data Fusion, Kalman Filtering, Bayesian Calibration, etc., etc. However, the statistical literature rarely shows applications to dynamic models (including PBMs), so for such modellers the applicability of the Bayesian perspective may be unclear. Here I hope to show how the multitude of presented techniques and approaches are in fact family relations of each other. The Bayesian perspective provides a united treatment of model-data issues, irrespective of the type of model or data.

2.4 Errors and Uncertainties in Modelling

The Bayesian approach is essential for scientists because it allows formal representation and comprehensive quantification of uncertainties as well as rigorous learning
from observations. Errors are unavoidable in modelling but we are uncertain about
their magnitude. Inputs (parameters, environmental drivers) can never be known perfectly, and every model is a simplification of reality and thus has structural errors. We
quantify our uncertainty about these errors in the form of probability distributions,
and propagate these uncertainties to model outputs, i.e. we assess how accurate and
precise our model predictions are. We use data and Bayes' Theorem to reduce the
uncertainties about model error. To be able to use Bayesian methods, we need to
have a clear understanding of the different kinds of error. So let's discuss them in
detail.

2.4.1 Errors and Uncertainties in Model Drivers

As described above, drivers are the environmental conditions that are input to a
model (the x in the Modelling Equation (2.2)). Methodologically, drivers are defined
as boundary conditions that we do not attempt to calibrate, but that we accept from
an external source of information. Although drivers are not calibrated, we are interested in any possible errors they may contain, and how uncertain we are about their
magnitude. Driver errors come in three types: gaps, measurement error and non-
representativeness, which each need to be treated in a different way.

Gaps in driver data are common. When a model requires daily driver data and the
weather time series misses some day or days, the gaps need to be filled. Gap-filling
methods range from simple linear interpolation to stochastic process modelling,
akin to kriging. The latter is the only approach for which uncertainties are readily
quantified (as the kriging or Gaussian process variance).

Measurement error is caused by limited precision and accuracy of the measurement instrument, and sometimes also by errors in transferring measured values to
data files. This is the simplest kind of error; uncertainty about it is generally well
represented as Gaussian noise.

The most complicated errors arise when the driver data are *not representative* of the
environmental conditions for which we plan to run our model. The weather data may
have come from an off-site weather station, or the data may have a different spatio-
temporal scale than the model requires. A very common example is using output
from gridded climate models as input for PBMs: the climatic data will then be grid
cell averages, thus ignoring spatial heterogeneity. Such data are not representative for
any specific ecosystem within the cell. If the model is nonlinear, as most models are,
then input averaging leads to errors in model outputs, but methods to estimate that
error exist, based on Taylor-expansion of the driver-dependency [Van Oijen (2017);
see Chap. 22]. Another option is to downscale the data, although that brings its

own errors and uncertainty. No general rule for assigning probability distributions to errors from non-representativeness exist: each case must be examined on its own.

2.4.2 Errors and Uncertainties in Model Parameters

Parameters are constants that represent fixed properties of our system of interest. In PBMs, every process that is modelled requires a parameter to define its basic rate. But process rates also depend on internal and external conditions (e.g. the nitrogen content of plants, or air temperature), and each controlling mechanism adds at least one sensitivity parameter to the model. The number of parameters in PBMs for vegetation modelling, for example, tends to range from several tenths to many hundreds (Cameron et al. 2013; Van Oijen et al. 2008, 2013). This will pose problems for parameter estimation as we shall discuss later.

Parameters can be set to incorrect values, making model behaviour unrealistic. But a precise definition of parameter error is hard to give. What is the 'true' value of a parameter? Because a model is a simplification of reality, every parameter plays a somewhat different role in the model than its namesake in the real world. We can therefore not simply go out and measure the true value exactly. This is apart from the fact that measurements have errors too and that no system property is ever truly constant in reality. In practice, we say that the correct value of a parameter is the value that makes the model behave most realistically. Parameter error is the difference from that value, and parameter uncertainty is not knowing what that value is.

Although measurements cannot provide true parameter values, they can give first rough estimates. For about thirty different parameters common to three forest PBMs, Levy et al. (2005) reviewed measurement values reported in the literature. For each parameter, a wide range of values was reported that could be interpreted as probability distributions representing parameter uncertainty. For many parameters, the distributions were skewed and better represented by beta distributions than by normal distributions (Van Oijen and Thomson 2010).

2.4.3 Errors and Uncertainties in Model Structure

Every model is a simplification of reality and therefore, to some extent, wrong. But the behaviour of models—their repertoire of outputs for different conditions—can be compared, showing that some models are more realistic than others. Comparisons of ecosystem models abound in the literature and tend to show large differences between their predictions (Cameron et al. 2013; Van Oijen et al. 2008). Some tentative general conclusions may be drawn: the feedback structure of models is more important than the mathematical form of individual equations (Hyvönen et al. 2007) and a consistent level of process detail in different parts of the model is desirable (Rollinson et al. 2017; Van Oijen et al. 2004).

Increasingly, comparisons of all models, including complex PBMs, involve assessing simulations against observations, not just of output variables such as productivity, but also of the underlying processes and mechanisms as represented in the models (Clark 2016; Milne et al. 2015; Johnson et al. 2016; Schlesinger et al. 2016). Medlyn et al. (2015) refer to this as an "assumption-centred model intercomparison approach". However, these model comparisons do not employ probability theory and therefore cannot quantify the degree of uncertainty about model structural error. Moreover, any advice about model structure based on observations is contingent on the range of environmental conditions for which the models were tested and remains heuristic: there is no unique way to derive a model from first principles, so errors remain inevitable.

This leaves us with only two ways to account, probabilistically, for structural error in a modelling study: extend the PBM with a stochastic error term (Kennedy and O'Hagan 2001), or use a large ensemble of different models and proceed as if one model in the set should be correct (e.g. Fu et al. 2012). In the first approach, uncertainty is quantified by assigning a probability distribution to the structural error term. In the second approach, uncertainty is represented by a probability distribution over the set of models, with highest probabilities assigned to models that are considered to be most plausible. Both require taking into account that model performance depends not only on its structure but also on the parameter settings with their own uncertainties. Technical details of both methods will be given in Chaps. 12 and 13.

2.4.4 Forward Propagation of Uncertainty to Model Outputs

Because measurements and models have unknown errors, all we can quantify are uncertainties. Therefore, the common term of 'error propagation' denoting how much the error in inputs contributes to error in outputs is a misnomer. What is propagated is uncertainty, not error. Outputs may have minimal error despite large errors in inputs, if the errors happen to have compensating effects. This is a common occurrence when models are tuned to produce a desired result.

The techniques, rather than the name, associated with 'error propagation' may well be used to quantify output uncertainty, provided the model is simple enough that partial derivatives of output with respect to inputs can be analytically calculated. However, PBMs tend to be too complex for such approaches, so uncertainty is mostly quantified by Monte Carlo methods: sampling from the probability distributions of model inputs (and structures if we have an ensemble of models) to generate a representative set of possible model outputs. In the study by Levy et al. (2005), mentioned above, Monte Carlo sampling was used to quantify the contribution of parameter and model structural uncertainty to uncertainty about the carbon sink of a coniferous forest in southern Sweden. They concluded that the carbon sink uncertainty was for 92% due to parameter uncertainty whereas model structural uncertainty accounted for only 8%. The key parameter uncertainties were for allocation of carbon to leaves, stems and roots. These results were of course contingent on the three chosen forest models and the single application site. Reyer et al. (2016) also demonstrated the importance of parameter uncertainty—in their case for prediction of future forest productivity—but in a comparison with uncertainty about climatic drivers rather than model structure. Sutton et al. (2008) showed the importance of uncertainties about model drivers (in particular atmospheric nitrogen deposition) and model structure for predictions of forest productivity across Europe. Minunno et al. (2013) found that uncertainty about soil conditions (water availability and fertility) mainly determined the predictive uncertainty of a growth model for *Eucalyptus globulus* in Portugal.

2.5   Bayes and Science

In all sciences, our aims are to explain observations and predict the future. For example, in environmental science and ecology, we use predictions to assess the likely impact of environmental change on ecosystems and to optimise management. The problems are very wide-ranging in data availability and model choice. Bayes can help us in all cases to process the information optimally. Irrespective of model structure, from the simplest linear regression to the most complex computer model, Bayesian methods can be used to quantify and reduce the modelling uncertainties. The two most basic uses of Bayesian methods are to tell us which parameter values and which models are plausible:

1. Parameter estimation $p[\theta|y]$, where θ is a scalar or a vector in parameter space.
2. Model evaluation $p[f_i|y]$, where $f_i \in \{f_j\}_{j=1..n}$.

Every science question can be put in the form: "What is $p[model|data]$?". That is just the scientific way of asking the question: 'What can observations teach us about the world?' So everything is about conditional probability distributions, and that is why probability theory, in yet more words of Jaynes (2003), is 'the logic of science'. In fact, R.T. Cox proved already in 1946 that the laws of probability theory can be derived from common sense ideas about consistent reasoning (Jaynes 2003). The world of Sherlock Holmes, with proofs rather than probabilities, and deduction rather than induction, is not science. Nothing can be unconditionally proven in science, there are always assumptions regarding the quality of our data and the quality of our models that force us to use probabilistic language.

2.6   Bayesian Parameter Estimation

We always have some uncertainty about the proper values of a model's parameters, and we express this uncertainty as the probability distribution $p[\theta]$. When new data y arrive, we want to use those data to reduce our uncertainty about θ. As we saw in the last chapter, Bayes' Theorem tells us how to change a *prior* parameter distribution, $p[\theta]$, into a posterior distribution for "θ given y" denoted as $p[\theta|y]$:

$$p[\theta|y] \propto p[\theta]p[y|\theta].$$

So according to Bayes' Theorem, the posterior distribution is proportional to the prior $p[\theta]$ and the likelihood $L[\theta] = p[y|\theta]$. Parameter estimation using this theorem is called 'Bayesian calibration' (Kennedy and O'Hagan 2001; Van Oijen et al. 2005). At this point, we briefly remind the reader that both the parameter vector θ and the data vector y can be (highly) multi-dimensional. In other words, the posterior distribution for the parameters, $p[\theta|y]$, is a *joint* probability distribution, and it is likely that most parameter-pairs in θ will show some degree of correlation. Let us assume, for now, that model error ϵ_{model} is much smaller than measurement error and can be ignored. We can then combine the Measurement and Modelling Equations ((2.1), (2.2)) to derive the likelihood function:

<u>LIKELIHOOD FUNCTION</u> (given $\epsilon_{model} = 0$) :

$$p[y|\theta] = p[f(x, \theta) - y = \epsilon_y].$$

(2.4)

If we would further assume that ϵ_y has a zero-mean normal distribution and a coefficient of variation of 50% (we used the same example before), then the likelihood function would simplify to $p[y|\theta] = N[f(x, \theta) - y; \mu = 0, \sigma = y/2]$. Say that we have also defined the prior distribution for the parameters, $p[\theta]$, based on literature review and expert opinion. Then the final step, as prescribed by Bayes' Theorem, is to find the product of the prior and the likelihood. In principle, a representative sample from the posterior distribution can easily be generated by Monte Carlo sampling: take a large sample from the prior and use the likelihoods as weights in deciding which parameter values to keep. More sophisticated methods such as MCMC will be explained in later chapters. This simple example captures all the essential steps in Bayesian calibration: (1) specify a prior, (2) specify $p[\epsilon_y]$ and from it the likelihood function, (3) apply Bayes' Theorem.

After a Bayesian analysis, we will have reduced our uncertainty about models and parameters. But we will of course primarily be interested in what we have learned about the real world itself. Such applied problems include forecasting, risk analysis, decision making. In each of those cases, we are interested in $p[z]$, which we estimate as $p[f(x, \theta) + \epsilon_{model}|y]$. The main paradigm shift that Bayesian thinking brings to science is that we no longer aim to find the single best parameter value or the single best model, instead we think in terms of probability distributions.

Our focus in the next few chapters will be on parameter estimation ($p[\theta|y]$). In later chapters, we shall consider model comparison or selection ($p[f|y]$).

Chapter 3
Assigning a Prior Distribution

There is no such thing as *the* prior probability distribution of a parameter or a set of models. A prior expresses uncertainty arising from incomplete knowledge, and whatever the subject is, people have different knowledge and expertise. So instead of speaking of "the prior probability of x", each of us should say "my prior probability for x". We *assign* a prior probability distribution, we do not *identify* it. This is even the case when we invite the opinion of experts on the likely values of our model's parameters. There is an art and science to eliciting expert opinion in order to formulate a Bayesian prior, and books have been written on the topic (see O'Hagan 2012), but ultimately the responsibility for the prior lies with the modeller, not the expert panel.

We do not have complete freedom in assigning just any distribution. First of all, we are working within probability theory, so we must make *coherent* probability assignments (Jaynes 2003; Lindley 1991). If we say that $p[A] = 0.5$, then we cannot say that $p[A \text{ or } B] = 0.4$. In general, if we have multiple parameters in our model, then our task a priori is to assign a valid *joint* probability distribution to all parameters.

Moreover, when you and I have exactly the same information about, say, a parameter θ, then we should be assigning the same probability distribution $p[\theta]$. When the parameter is by definition non-negative, as most physical properties are, then a probability distribution such as the Gaussian which ranges from minus to plus infinity, is inappropriate. However, we need not be too principled here. A Gaussian distribution $N[\mu, \sigma^2]$ with positive mean μ and a typical coefficient of variation σ/μ of 20%, will have less than one-millionth of its probability mass below zero. Also, when you are expecting many data to be forthcoming, then the role of the prior becomes small: the posterior distribution will then be mainly determined by the likelihood function. However, there is a tendency among researchers to select too wide, uninformative prior distributions. An extreme example of this is *maximum likelihood estimation* (MLE) which does not require you to specify any prior distribution at all, and behind the scenes is assuming that any parameter value is a priori equally plausible. This is bad use of information and should generally be avoided. If you have enough information about a system to build a model for it, with parameters that you have given

© Springer Nature Switzerland AG 2020
M. van Oijen, *Bayesian Compendium*,
https://doi.org/10.1007/978-3-030-55897-0_3

implicit meaning by the role they play in your model, then it is not plausible to have no idea about parameter values at all. As a simple example, if you believe that $f(x, \theta) = x\theta$ is a proper model for your system, then you are saying that output $f()$ responds linearly to x. When could you make such a statement without having any idea about whether the proportionality is negative, zero, or positive?

3.1 Quantifying Uncertainty and MaxEnt

To avoid the opposite extremes of ignoring parameter uncertainty and assuming it infinite, the general principle in assigning priors should be to choose that distribution which expresses maximum uncertainty while being constrained by whatever information you have. But that raises the question: what is the appropriate quantitative measure of uncertainty? We can immediately rule our the *variance*, as a simple example will show. Take a fair die, such that our uncertainty about the number of dots showing after the next throw is uniformly distributed as $p[i] = \frac{1}{6}; i \in \{1..6\}$. That complete uncertainty has a variance of 2.92, whereas a much less uncertain distribution with $p[1] = p[6] = 0.5$ and $p[i \in \{2, 3, 4, 5\}] = 0$ has the maximum possible variance for a six-sided die of 6.25. A much better measure of uncertainty is the *entropy S* of the distribution, defined for discrete probability distributions as:

ENTROPY OF DISCRETE DISTRIBUTION:

$$S = -\sum_{i=1}^{n_p} p[i] ln(p[i]), \tag{3.1}$$

and a similar equation exists for continuous distributions with an integral replacing the summation. Just like Cox showed formally that consistent rational reasoning requires us to follow the rules of probability theory, Jaynes (2003) showed that for uncertainty we should use entropy. His *Maximum Entropy* (MaxEnt) principle for assigning probability distributions has become a standard approach in many areas of science, and it has even been implemented in an R-package called FD by Lalibert and Shipley. Let's try it out on the example of the fair die that we have just given. We ask for the discrete probability distribution over $\{1 : 6\}$ with maximum entropy under the constraint that the mean is equal to 3.5. That can be answered with the following very short R-code:

```
Dp  <- 1:6 ; meA <- FD::maxent(3.5, Dp)
pA <- meA$prob ; SA <- meA$entropy
```

This code gives pA as the MaxEnt distribution, which we plot in panel A of Fig. 3.1. It is, as we expected, the uniform distribution which has entropy equal to SA = 1.792.

The other panels in Fig. 3.1 show the MaxEnt distributions for when we have different information. Panel B assumes a mean of 4.5 which makes the MaxEnt

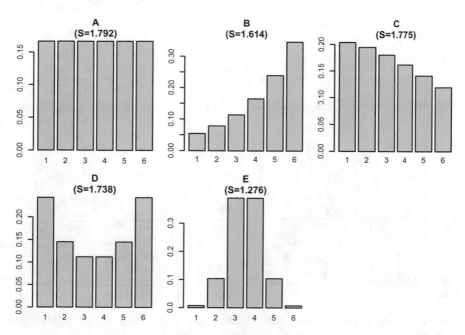

Fig. 3.1 Distributions selected using MaxEnt. Entropies between brackets. All distributions are on the same domain (1,..,6) but with different constraints for mean and mean square: see text

distribution slope upward. Panel C assumes a mean value for i^2 of 13. Panel D assumes a mean for i of 3.5 and for i^2 of 16. Panel E is as D but with mean square 13 instead of 16.

We show two more MaxEnt distributions in Fig. 3.2. The panel marked F has the same constraints as panel E in the first figure, but we have refined the domain from $\{1, .., 6\}$ to $\{1.0, 1.1, .., 5.9, 6.0\}$. So we have gone from a six-sided die to a 51-sided one. You see that the distribution then becomes far more bell-shaped. In the limit of a continuous distribution with prescribed mean and mean-square (or variance), MaxEnt identifies a Gaussian distribution as the most uncertain one. So if all you learn about a parameter is the first two *moments* of its distribution, MaxEnt suggests that you assign a Gaussian distribution. Note further that entropies cannot be compared for distributions over different domains: entropy values S are higher for distribution F than for E only because of its greater domain multiplicity.

The final panel G shows an elegant result for a 10-sided die. The depicted distribution was identified by MaxEnt given constraints for the mean of i and the mean of i^3 (third moment), and this produced the beautiful wave pattern as the most uncertain distribution. I leave you to verify for yourself that the distributions in the two figures have maximum entropy within their different constraints, and perhaps explore some more distributions of your own.

There is a rich literature on maximum entropy and related principles for assigning prior probability distributions, and Jaynes is a very good place to start, as is the older

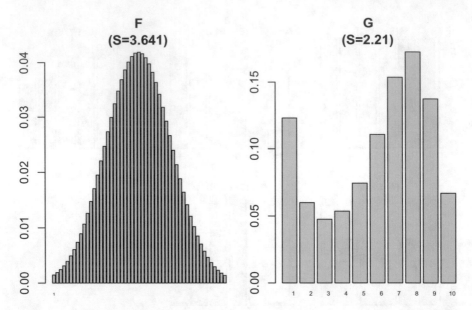

Fig. 3.2 More MaxEnt distributions. Domains cover 51 and 10 possible values. Entropies between brackets

work of Jeffreys (see e.g. Robert et al. 2009). An important special class of parameters identified by Jeffreys is that of *scaling* parameters, such as standard deviations σ and variances σ^2. These are non-negative parameters that indicate the scale over which other variables vary, and Jeffreys and Jaynes showed that, without other knowledge, the prior density for such parameters should decrease inversely with their magnitude, i.e. $p[\sigma] \propto \frac{1}{\sigma}$ defined on the whole domain from zero to plus infinity. This is called a *Jeffreys Prior*. However, we often do have more information about the variance than just non-negativity, and then the Jeffreys Prior will not be the right choice.

3.2 Final Remarks for Priors

Bayes is ubiquitous: the same Bayesian methodology is used throughout the sciences. However, practical problems can vary between different types of application. In disciplines where prior information is scarce, say in the study of rare species, we can only assign wide, uninformative prior distributions, and we need to be very careful about what the data tell us about the relative likelihoods of different parameter values. In other disciplines, say agriculture, there may be rich prior information but from disparate sources, making it difficult to come up with a proper prior distribution. It may also be the case that you are modelling the behaviour of a population of organisms, or variables that are dispersed over different locations. In such cases, a *hierarchical prior* may be useful as we shall discuss in Chap. 16.

Whatever the circumstances, in most cases there will be a degree of apparent subjectivity in the choice of prior, because people have different background knowledge. Therefore, in publications, it is important to state clearly what priors you have chosen, and based on what information. It will also be a service to the reader if you show how much the results of your Bayesian calibration change when choosing a different prior.

The next chapter considers the second element in Bayes' Theorem: the likelihood function.

Chapter 4
Assigning a Likelihood Function

As scientists, we want to know how to parameterise our models, make comparisons with other models, and quantify model predictive uncertainty. For all these purposes, measurement data are needed, but how exactly should we use the data? The answer is always the same: in the *likelihood function*. The likelihood is a conditional probability density function, denoted as $p[y|\theta]$ (or more succinctly as $L[\theta]$). It is the answer to the question: "what is the probability of measuring y if the true value is $f(x, \theta) + \epsilon_{model}$?". This can be written formally as follows:

LIKELIHOOD FUNCTION:
$$L[\theta] = p[y|\theta] = p[f(x, \theta) - y = \epsilon_y + \epsilon_{model}]. \tag{4.1}$$

This definition of the likelihood function follows from the Science Equation (Eq. (2.3)) that we presented in Chap. 2. In that same chapter we already gave a simplified definition of the likelihood function where ϵ_{model} was assumed to be zero. We shall largely ignore model structural error in this chapter as well, but return to that issue in Chaps. 12 and 13.

We use the likelihood function, together with the prior distribution discussed in the preceding chapter, in Bayes' Theorem to derive the posterior distribution for our parameters. But, just as with the prior distribution, we do not *derive* the likelihood function, we have to *define* it ourselves. So there is an element of subjectivity here: our choice of likelihood function will depend on what we believe the data can tell us. And that will depend on our views of the possible errors in the data. I would say that in practice, the hardest and most important step in Bayesian calibration is formulating an appropriate likelihood function.

The likelihood is a powerful concept: y is usually multi-dimensional and may consist of measurements over time on several different soil and tree variables, yet $p[y|\theta]$ can always be defined. It is through the likelihood function that Bayesian calibration has the capacity to use highly heterogeneous data sets in parameter estimation. Levy et al. (2017) used Bayesian calibration to reconcile eddy-covariance

© Springer Nature Switzerland AG 2020

M. van Oijen, *Bayesian Compendium*,

https://doi.org/10.1007/978-3-030-55897-0_4

measurements of N_2O emissions with chamber measurements of the same fluxes. Patenaude et al. (2008) combined remote sensing data from satellites with field-based data on Corsican pine stands in the UK, in Bayesian calibration for the parameters of forest PBM 3-PG. Höglind et al. (2016) combined measurements on 10 different variables from 5 grassland sites in their likelihood function. With such rich data sets, it becomes important to assess whether measurement errors for different variables have correlations that should be expressed in the likelihood function.

Formulating the likelihood function can be difficult, even when model error ϵ_{model} can be ignored and different variables are measured independently, because measurements can be wrong in three different ways, as summarized in the next equation:

$$\epsilon_y = \epsilon_{y,stochastic} + \epsilon_{y,systematic} + \epsilon_{y,representativeness}. \tag{4.2}$$

The first of the three data error terms, $\epsilon_{y,stochastic}$, quantifies random measurement noise, independent for each data point in y, which we could represent with a zero-mean normal distribution (Ogle and Barber 2008). This is often the only data error recognized by modellers, but it is not always the most important one. The second error term, $\epsilon_{y,systematic}$ represents measurement bias which could shift whole collections of data up- or downward (beautifully illustrated by Jaynes (2003, pp. 257–260) with his emperor of China parable). The final term, $\epsilon_{y,representativeness}$, is generally the hardest to quantify. It refers to the possibility of our data being derived from other conditions than our model is designed for. Take the example of a model for crop growth. If the observed crop has a hidden growth limitation, say phosphorus deficiency, that is not expressed in the model, then the data—from the point of view of model parameter estimation—will be underestimating growth. In contrast to the other two types of measurement error, the scope for $\epsilon_{y,representativeness}$ cannot be reduced by more careful measurement or greater sampling intensity.

For every data set, the modeller and data expert need to distinguish the three types of possible data error, and assign probability distributions to the respective error uncertainties. The parameters of these probability distributions, such as the standard deviation of the stochastic noise, can be fully specified a priori (e.g. $\sigma_{y,stochastic} = y/2$), or they can be included with the other parameters to be estimated in the calibration. The latter method would be preferable in this example if we have little information about the precision of our measurement instrument. Van Oijen et al. (2011) calibrated the degree of systematic error in chamber-measurements of soil-fluxes of CO_2, N_2O and NO using four different forest PBMs—and all four calibrations suggested that the measured CO_2 emission values had been unrealistically high.

4.1 Expressing Knowledge About Data Error in the Likelihood Function

When calibrating their model, many modellers as mentioned assume that their model is correct and that their uncertainty about every data point can be described by a Gaussian distribution with constant variance σ_y^2. Their likelihood function is then:

$$p[y|\theta, f] = |2\pi \Sigma_y|^{-1/2} \exp[-\frac{1}{2}(f(\theta) - y)^\top \Sigma_y^{-1}(f(\theta) - y)], \qquad (4.3)$$

where the covariance matrix has σ_y^2 everywhere on the diagonal and zeros elsewhere: $\Sigma_y = \sigma_y^2 \mathbf{I}_{n_y}$. That commonly used Gaussian function presumes that all measurement errors $\{y_i\}, i = 1, .., n_y$ are independent from each other, and that measurement uncertainty is the same irrespective of the value of y. But those assumptions generally do not do justice to everything we know. We often have information about the measurement process that we can include in the covariance matrix Σ_y. One important example is that spatially or temporally nearby observations may have similar measurement error. Our measurement instrument may have drift and conditions for measurement may be harder in certain areas than in others. We can easily express such error correlations in the off-diagonal terms of Σ_y (see e.g. Gregory 2005) using ideas from Gaussian Process modelling (Chap. 14) and time series analysis (Chap. 21).

It will usually also be better not to have the exact same value everywhere on the diagonal of Σ_y because measurement error is likely to vary with the magnitude of the y_i. Estimating the weight of an elephant comes with greater uncertainty variance than estimating the weight of a mouse. So our uncertainty about y increases with its magnitude, but how strongly? The opposite extreme to setting the variance σ_y^2 as a constant is a constant coefficient of variation ($c.v. = \sigma/y$). That is equivalent to the variance being proportional to y^2, but that may underestimate our uncertainty for small y. A more reasonable intermediate position is to have our variance be proportional to y.

Let's compare these three approaches in a concrete example. We have data (x, y) with $x = (1, 2, 3)$ and $y = (10, 1, 5)$, and want to fit a straight line. Let's assume that we have no prior knowledge at all about the slope and intercept—which is of course unrealistic but serves to highlight the role of the likelihood in the calibration. Let's define the three Gaussian likelihoods as above, with always a variance of 1 for the first measurement point: $\sigma_y^2[1] = \Sigma_y[1, 1] = 1$. For the two other points we have our different choices:

1. variance independent of y: $\Sigma_y[2, 2] = \Sigma_y[3, 3] = 1$,
2. variance proportional to y: $\Sigma_y[2, 2] = 0.1$; $\Sigma_y[3, 3] = 0.5$,
3. variance proportional to y^2: $\Sigma_y[2, 2] = 0.01$; $\Sigma_y[3, 3] = 0.25$.

We do the Bayesian calibration for these three choices and that gives us the following posterior distributions $N[\mu_{\beta|y}, \Sigma_{\beta|y}]$ for the parameter vector $\beta = (slope,$

intercept). [We leave the mechanics of how to find that posterior to the following chapters.]

CALIBRATION WITH 3 LIKELIHOOD FUNCTIONS:

$$\text{Case 1: } \mu_{\beta|y} = \begin{bmatrix} 10.3333 \\ -2.5 \end{bmatrix}; \quad \Sigma_{\beta|y} = \begin{bmatrix} 2.333 & -1 \\ -1 & 0.5 \end{bmatrix}.$$

$$\text{Case 2: } \mu_{\beta|y} = \begin{bmatrix} 3.9474 \\ -0.7895 \end{bmatrix}; \quad \Sigma_{\beta|y} = \begin{bmatrix} 1.553 & -0.711 \\ -0.711 & 0.342 \end{bmatrix}. \quad (4.4)$$

$$\text{Case 3: } \mu_{\beta|y} = \begin{bmatrix} -1.3566 \\ 1.2791 \end{bmatrix}; \quad \Sigma_{\beta|y} = \begin{bmatrix} 0.847 & -0.413 \\ -0.413 & 0.203 \end{bmatrix}.$$

So we see very large differences between the distributions! In Fig. 4.1, we show the three lines that are represented by the posterior mean parameter vectors. The figure shows that with likelihood function 1, the first and highest data point is very influential, whereas that point is largely ignored in calibration 3. Case 2 may be the happy intermediate, but it depends of course on your knowledge about the measurement process.

Note that you can always decide to estimate the σ_y^2 in your calibration rather than prescribe them as we did here. That approach may be a wise choice if you have little prior knowledge about the variability of measurement error. But you then obviously need to assign a joint prior probability distribution to the $\{\sigma_y^2(i)\}$, and should try to include in that prior whatever you can say about the measurement error you expect.

We already mentioned that we can represent correlated uncertainties about data points in the off-diagonal terms of the covariance matrix. That is a specific example of a more general important consideration when formulating our likelihood function: we should not exclusively focus on the values of individual data points, but evaluate

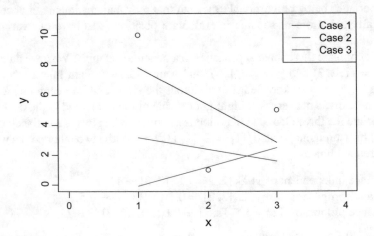

Fig. 4.1 Posterior mean straight lines derived by Bayesian calibration using three different likelihood functions

patterns of system behaviour. For example, if the values of y show an oscillating pattern over time, then parameter vectors that instead lead to stabilising behaviour should be accorded low likelihood. Spectral analysis of y and analysis of statistical moments may help us quantify patterns succinctly.

4.2 What to Measure

This chapter has discussed the likelihood function as the means by which empirical information enters into our analysis. But we have not discussed what data to gather in the first place as that is for the most part too problem-specific an issue. But you can use the Bayesian approach to evaluate different measurement strategies. For example, when considering to replace a piece of equipment with a higher-precision tool that is more expensive, we could first run Bayesian calibrations with virtual data of differing precision in order to assess how much the new tool is likely to reduce predictive uncertainty (Van Oijen et al. 2005). And during the course of your experiments and monitoring studies, you may want to change your measurement set-up if it no longer gives you the information that you need. That is the topic of Bayesian adaptive sampling which will be discussed in Chap. 23.

Chapter 5
Deriving the Posterior Distribution

In the preceding chapters, we discussed how we can assign a prior distribution for our parameters, and how to choose a likelihood function that captures the information content of our data. So all that is left, is to apply Bayes' Theorem (Eq. (1.2)) to derive our desired posterior distribution. Note that when talking about the posterior, we use the phrase 'deriving the' distribution rather than 'assigning a' distribution. That is because Bayes' Theorem tells us exactly what the posterior distribution should be, once we have defined our prior and likelihood.

Even though Bayes' Theorem is simple to write down, the right-hand side is not in a convenient form: it is often not easy to see what kind of probability distribution is defined by the product of prior and likelihood. So how do we use Bayes' Theorem in practice? Well, there are two ways of using the theorem to find the posterior distribution: *analytically* and *numerically*. The analytical method is the truly mathematical way, it consists of writing down the formula for the posterior distribution directly. But this can only be done in certain simple cases, e.g. when both the prior and the likelihood are Gaussian distributions. The posterior distribution is then also Gaussian. There are a few other combinations of probability distributions that allow you to directly write down the posterior distribution (which then always belongs to the same type as the prior), but we shall not study many of such so-called 'conjugate' distribution pairs in this book. They are generally not good reflections of the information that we have a priori or receive in new data. However, we should be aware of these methods because they are implicitly used by standard linear models, so we do work out an example in the following section.

5.1 Analytically Solving Bayes' Theorem: Conjugacy

Bayes' Theorem says that the posterior is proportional to prior times likelihood. As mentioned, we can calculate the posterior analytically if the prior and likelihood belong to conjugate distributions. A simple example is when we want to estimate a scalar parameter for which both prior and likelihood are univariate Gaussians. Then

© Springer Nature Switzerland AG 2020
M. van Oijen, *Bayesian Compendium*,
https://doi.org/10.1007/978-3-030-55897-0_5

the posterior will also be a univariate Gaussian. Say we have the following definitions for prior $p[\theta]$ and likelihood $p[y|\theta]$ where the dataset y consists of n measurements y_i of the parameter, each with the same measurement uncertainty σ_y^2:

$$p[\theta] = \frac{1}{\sigma_0\sqrt{2\pi}} \exp\left[-\frac{1}{2}\left(\frac{\theta - \mu_0}{\sigma_0}\right)^2\right], \tag{5.1}$$

$$p[y|\theta] = \prod_{i=1}^{n} \frac{1}{\sigma_y\sqrt{2\pi}} \exp\left[-\frac{1}{2}\left(\frac{y_i - \theta}{\sigma_y}\right)^2\right]. \tag{5.2}$$

In that simple case, the posterior distribution $p[\theta|y]$ is:

$$p[\theta|y] = \frac{1}{\sigma_1\sqrt{2\pi}} \exp\left[-\frac{1}{2}\left(\frac{\theta - \mu_1}{\sigma_1}\right)^2\right], \tag{5.3}$$

where

$$\mu_1 = \frac{1/\sigma_0^2}{1/\sigma_0^2 + n/\sigma_y^2}\mu_0 + \frac{n/\sigma_y^2}{1/\sigma_0^2 + n/\sigma_y^2}\bar{y},$$

$$\frac{1}{\sigma_1^2} = \frac{1}{\sigma_0^2} + \frac{n}{\sigma_y^2},$$

$$\bar{y} = \frac{1}{n}\sum_i y_i.$$

Let's make this example concrete by assuming some values for the data and the prior. We choose data $y = (1.0, 0.9, 1.1)$ with variance $Vy = 1$, and for the prior we choose a mean $m0 = 0$ and a variance $V0 = 1$. This is the R-code for these choices that then calculates the posterior:

```
m0    <- 0              ; V0 <- 1
y     <- c(1.0,0.9,1.1) ; ny <- length(y) ; my <- mean(y) ; Vy <- 1
k     <- (1/V0) / (1/V0 + ny*Vy)
m1.a <- k * m0 + (1-k) * my ; V1.a <- 1 / (1/V0 + ny/Vy)
```

When we run this code, we find that the posterior mean is m1.a = 0.75 and the posterior variance is V1.a = 0.25. You can verify that that is consistent with the formulas shown above. In the next section, we show how to estimate the posterior for this problem numerically.

5.2 Numerically 'Solving' Bayes' Theorem: Sampling-Based Methods

When we cannot find the posterior analytically, we have to use numerical methods, and that is the usual case. In the example of this chapter, an analytical solution was of course possible, but let's tackle the problem numerically anyway. The numerical approaches always involve *sampling*. They do not give us a nice formula for the posterior distribution but a representative sample from it. There are many different algorithms for numerically sampling from a posterior distribution, but here we are going to use the simplest one, which is called the *Accept-Reject* (A-R) algorithm or *rejection sampling*. The A-R algorithm begins with a large sample from the prior which it then thins out to produce a representative sample from the posterior. It works as follows:

1. Calculate the largest possible likelihood value L_{max}.
2. Take a large sample $\{\theta_i \; ; \; i = 1, .., n\}$ from the prior distribution for the parameters.
3. For each parameter vector, draw a random number from the standard uniform distribution $u_i \sim U[0, 1]$ and accept θ_i in the final sample if $u < L[\theta_i]/L_{max}$.

Let's implement this in R, using $n = 10^5$, and do the sampling:

```
Lmax          <- prod( dnorm( y, my, sqrt(Vy) ) )
n             <- 1e5 ; samplePrior <- rnorm( n, m0, sqrt(V0) )
L             <- function(b){ prod( dnorm( y, b, sqrt(Vy) ) ) }
Lrelative     <- sapply(samplePrior,L) / Lmax
iAccept       <- which( runif(n) < Lrelative )
samplePost <- samplePrior[ iAccept ]
m1.n          <- mean( samplePost ) ; V1.n <- var( samplePost )
```

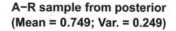

**A–R sample from posterior
(Mean = 0.749; Var. = 0.249)**

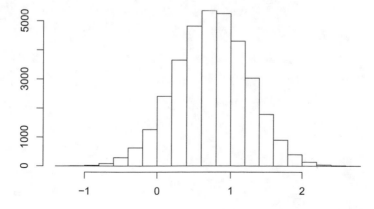

Fig. 5.1 Sampling from the posterior distribution of this chapter by means of Accept-Reject

The results of the A-R sampling are shown in Fig. 5.1. We see that we get nearly the same results as the analytical solution—there obviously is some sampling error. The A-R algorithm is not very good, and should only be used for very simple, low-dimensional problems. Imagine sampling from a 20-dimensional parameter space with the A-R algorithm. To sample each 'corner' of that space just once would already require a sample size of $n = 2^{20} \approx 10^6$. There are modifications of the A-R algorithm, such as *Importance Sampling*, which do not have the wasteful rejection step, but for efficiency we should really look at a different class of numerical distribution sampling methods called *Markov Chain Monte Carlo* (MCMC) methods, and that is the subject of the next two chapters.

Exercises

1. Inference from the analytical formula. The analytical solution of Bayes' Theorem (Eq. (5.3)) can make us quickly check certain general properties of Bayesian thinking.

- If you set n to zero, what do you get?
- If you take the limit of infinite data availability ($n \to \infty$), what do you get?

2. Explore the A-R algorithm. A downside of the A-R algorithm is that it requires a very large initial sample of size n from the prior. Rerun the A-R code with smaller values of n to see at which point sampling error becomes large (say > 1%).

Chapter 6
Sampling from Any Distribution by MCMC

6.1 MCMC

The Bayesian approach to parameter estimation requires modellers to make a major mental shift: we no longer aim to find a single 'best' parameter vector—instead we aim to determine the posterior probability distribution for the parameters. Only the full probability distribution adequately represents our state of knowledge. Although this shift has made rigorous uncertainty quantification possible, it has also created computational problems: when models have many parameters, the distribution will be high-dimensional and difficult to sample from. A solution for this problem was provided by Metropolis et al. (1953) who introduced the so-called Markov Chain Monte Carlo (MCMC) method. MCMC is the workhorse of computational Bayesian statistics, and by now many different MCMC algorithms exist.

Sampling always comes with error. But MCMC algorithms are guaranteed to converge—if we keep increasing the sample size—to the true distribution. That means that the properties of our sample (mean, mode, median, variance etc.) will get arbitrarily close to those of the true distribution, provided we keep sampling long enough. A Bayesian will mainly use MCMC to generate a representative sample from the posterior distribution, but MCMC can be used to sample from any probability distribution you may be thinking of, as we shall show below.

MCMC consists of a walk through parameter space in such a way that the visited points together form a representative sample from the distribution that we are after. The method of MCMC requires that at each step we decide to accept or reject a proposed new point (= vector of parameter values) depending on the prior probability times the likelihood for that parameter vector. Many different variants of MCMC now exist following that general plan. A simple introduction to the method, with an example for a process-based vegetation model, was provided by Van Oijen et al. (2005), but you will find explanations of MCMC in nearly every Bayesian text.

MCMC is more efficient than methods such as Accept-Reject that we looked at in the preceding chapter because it does not explore the whole parameter space. Instead it focuses the sampling effort on the region of highest posterior probability;

© Springer Nature Switzerland AG 2020
M. van Oijen, *Bayesian Compendium*,
https://doi.org/10.1007/978-3-030-55897-0_6

in most circumstances a chain of 10^4 to 10^5 steps is long enough even for quite parameter-rich models. However, MCMC does require evaluation of the model for each new proposed parameter vector in order to calculate the likelihood. And if the data that are compared with model output in the likelihood function are derived from n different sites (each with different drivers x), then we need to run our model n times at each step. So computational demand still can be considerable, all the more so for PBMs that are run for long time series. One proposed solution has been to include only a fraction of the model's parameters in the calibration (e.g. Minunno et al. 2013), a practice referred to as *parameter screening*. Parameter screening speeds up calibration by reducing the dimensionality of the parameter space that the MCMC needs to traverse. However, it thereby underestimates parameter uncertainty and should if possible be avoided.

It is high time to show an MCMC algorithm in more detail, so let us study some code.

6.2 MCMC in Two Lines of R-Code

To study the general properties of MCMC-algorithms, we start with sampling from a *Bernoulli distribution*, which is a univariate distribution for a parameter b, where the parameter can only assume two possible values: 0 or 1. This is the simplest possible probability distribution, and you could use it to model the throw of a coin. We choose the Bernoulli distribution with even odds, which is denoted as a Br[0.5] distribution, and sample from it with the following two lines of code:

```
nb   <- 1e1 ; b   <- rep(NA,nb ) ; b[1] <- 0 ; pjump <- 0.5
for(i in 2:nb) { jump <- (runif(1)<pjump) ; b[i] <- (b[i-1]+jump)%%2 }
```

We plot the progress of the sampling in the top two panels of Fig. 6.1. This is our first MCMC in action! The parameter space is extremely simple here, it consists of just the two points 0 and 1. The characteristic "walk through parameter space" of MCMCs is shown on the left in the so-called *trace plot*. The walk consists of short sequences of one or more 0's alternating with one or more 1's. We can also see that our chain is too short: our sample from the distribution consists of seven 0's and three 1's whereas a representative sample from a Br[0.5] distribution should have equal numbers of 0's and 1's. In MCMC jargon, we say that the chain has not yet *converged*. But our chain *mixes* well, it does not get stuck in one part of parameter space. And if we run the chain for 1000 iterations, we do get nearly equal numbers of 0's and 1's, as the bottom two panels of Fig. 6.1 show.

Our MCMC mixes well because we have chosen a good *proposal distribution*, which is the probability distribution that is applied at each iteration to determine where the chain will try to go next. In our code, the proposal distribution is actually also a Br[0.5], which means that the chain will 50% of the time propose to visit the 'other' point in parameter space, and in this particular MCMC that proposal is always accepted. [That makes our MCMC an example of a so-called

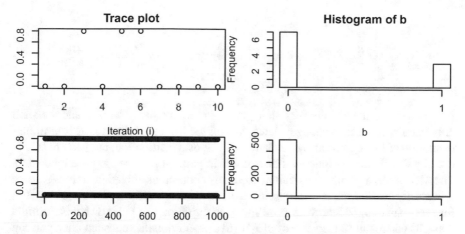

Fig. 6.1 Sampling from a Bernoulli distribution using MCMC. Top panels: MCMC with 10 iterations; bottom panels: 1000 iterations. Left: trace plots; right: chain histograms

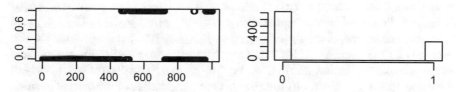

Fig. 6.2 Sampling from a Bernoulli distribution using a poorly mixing (jump probability 0.01) MCMC with 1000 iterations. Left: trace plot; right: chain histogram

Bernoulli stochastic process that could also be produced in basic R with the command `rbinom(1000,1,0.5)`.]

So what do you think would happen if we had chosen a Br[0.01] proposal distribution, with only 1% jump probability at each iteration of the MCMC? The results are shown in Fig. 6.2, where you can see that the algorithm now tends to get stuck for long sequences in either the 0 or the 1 position. In other words, mixing is poor and after 1000 iterations we have clearly not yet converged. But if you would continue this poorly mixing chain long enough (say for 10^5 iterations), you would finally see near-equal proportions of 0's and 1's. So even poorly mixing MCMCs are guaranteed to converge in the end, which is one of the reassuring properties of MCMC.

6.3 The Metropolis Algorithm

Now we come to a more full-fledged MCMC-algorithm which was proposed by Metropolis et al. (1953) nearly seventy years ago, but has only found widespread application over the past three decades, with the advent of fast computers. Let's jump right in and show you the R-code that implements the algorithm:

```
Metropolis <- function( p, b0=mb, SProp=Sb/100, ni=1e4 ) {
  bChain <- matrix( NA, nrow=ni, ncol=length(b0) ) ; bChain[1,] <- b0
  for( i in 2:ni ) {
    b1 <- bChain[i-1,] + as.numeric( rmvnorm(1,sigma=SProp) )
    if( runif(1) < p(b1)/p(b0) ) bChain[i,] <- b0 <- b1
    else                         bChain[i,] <- b0 }
  return(bChain) }
```

You see that the code is very short, but it is remarkably powerful, and we shall only marginally need to change it later. As you see from the 'rmvnorm', our implementation of the algorithm uses a Gaussian proposal distribution, but you are free to use any distribution as long as it does not skip parts of parameter space by design. The Metropolis algorithm can sample from any continuous univariate or multivariate probability distribution p as long as it can calculate the probability ratio $p[b1]/p[b0]$ for every pair of parameter vectors $b0$ and $b1$ in the domain of p. If you have a formula that tells you for any b the value of $p[b]$ up to a proportionality constant, then you can of course calculate every ratio $p[b1]/p[b0]$ because the proportionality constant then always drops out. The Metropolis algorithm can be described in one sentence as 'a walk through parameter space where at each step a candidate new position $b_{candidate}$ is generated randomly and accepted with probability $p[b_{candidate}]/p[b_{current}]$ (and with guaranteed acceptance if $p[b_{candidate}] > p[b_{current}]$)'. That single sentence also accounts for the meaning of 'MCMC': the 'Monte Carlo' refers to the randomness of the proposal generation, and the 'Markov' refers to the fact that the past history of the chain does not matter for its future behaviour, knowledge of the current position is always enough.

Let's make this concrete with an example. Say we want a sample from an unknown univariate probability distribution $p[\theta]$ on $[-\infty, \infty]$ where all we know is that the probability decreases exponentially with θ squared. In other words, $p[\theta] \propto exp(-\theta^2)$. Now provide a representative sample of magnitude 10^4 from this distribution using the Metropolis algorithm. Well, we already implemented the algorithm above, so we can do this with only one more line of R-code:

```
fA <- function(b){exp(-b^2)} ; bA <- Metropolis( fA, b0=0, SProp=diag(1) )
```

And we show the results in Fig. 6.3.

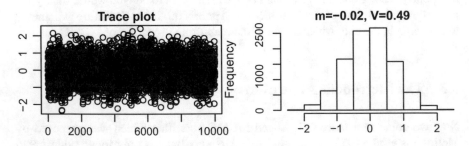

Fig. 6.3 Sampling from an unknown distribution using the Metropolis algorithm, where p[b] is proportional to the exponential of minus b squared

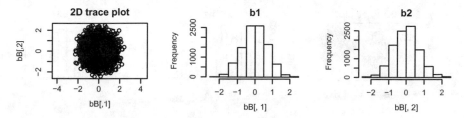

Fig. 6.4 Sampling from an unknown bivariate distribution using the Metropolis algorithm (see text)

In this case, you could instead of MCMC have used basic R to sample from the distribution: `bdirect <- rnorm(1e4,0,sqrt(2)/2)`. In other words, our Metropolis sampled from a $N[0, \sqrt{2}/2]$ distribution.

Let's now show that the same Metropolis algorithm can also sample from unknown multivariate distributions. This time our $p(\theta)$ is a bivariate distribution on the real number plane, with $p(\theta) \propto exp(-\theta[1]^2 - \theta[2]^2)$. Our code thus is:

```
fB <- function(b) { exp(-b[1]^2 - b[2]^2) }
bB <- Metropolis( fB, b0=c(0,0), SProp=diag(2) )
```

And this produces the results of Fig. 6.4.

And also this distribution could have been sampled with an existing R-function: `rmvnorm( 1e4, c(0,0), diag(2)/2 )`. Do you recognize the distribution? You need to be careful here because of the idiosyncrasies of R. The multivariate Gaussian sampling function 'rmvnorm' requires the user to specify parameter variances and covariances   in the form of the covariance matrix—whereas the equivalent function for univariate Gaussians 'rnorm' requires the user to specify the standard deviation, not the variance! [Even worse, later we shall encounter R-software that uses the *inverse* of the variance ...]

Our final example will show that the Metropolis algorithm will also sample correctly from multimodal distributions. This time what we know about our unknown probability distribution is the following:

- $p(\theta)$: bivariate, disjoint, unknown distribution on the real number plane
- $p(\theta) = 1$ if distance from $(0, 0)$ is less than 1, and if the distance from $(2, 1/2)$ is less than $1/2$. Otherwise $p(\theta) = 0$.

We code this in the usual way and show the results in 6.5.

```
fC <- function(b) { b[1]^2 + b[2]^2 < 1 || (b[1]-2)^2 + (b[2]-0.5)^2 < 0.5 }
bC <- Metropolis( fC, b0=c(0,0), SProp=diag(10,2) )
```

We see that the Metropolis algorithm was able to bridge the gap of zero probability between the two modes of the distribution. Both modes are sampled because we used a proposal distribution that very occasionally will propose a jump to the other mode. So in this case, a proposal distribution that never makes long jumps would have invalidated the algorithm.

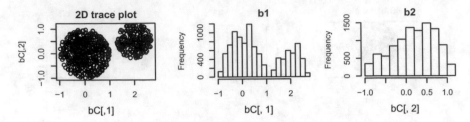

Fig. 6.5 Sampling from an unknown bimodal, bivariate distribution using the Metropolis algorithm (see text)

Exercise

1. Metropolis for sampling from the posterior. Take the A-R example at the end of the last chapter, of which the results were shown in Fig. 5.1. Can you replace the A-R with the Metropolis algorithm to sample from the desired posterior distribution? How would the prior and likelihood appear in the Metropolis? [This question makes you think ahead about what will be explained in the next chapter.]

Chapter 7
Sampling from the Posterior Distribution by MCMC

7.1 MCMC and Bayes

In the examples of MCMC in the preceding chapter, no prior or likelihood was specified, nor was there any talk of a posterior distribution. So why is MCMC important for Bayesian analysis?

We have seen that MCMC is a method for sampling from a probability distribution. And as Bayesians we always want to find a specific probability distribution, namely the posterior distribution. Moreover, when trying to find the posterior distribution, we are often in the situation, required for MCMC, of being able to specify the ratio of probabilities of two parameter vectors, without being able to directly quantify the probability density of any individual parameter vector. That is because Bayes' Theorem tells us that the posterior density for a parameter vector is proportional to the prior density for that vector times its likelihood. So now let's start using MCMC in earnest, for Bayesian calibration of the parameters of a model.

7.1.1 MCMC and Models

Our data are almost always in the form (x, y) where the y are observations and the x are the corresponding conditions such as covariates, coordinates, or times. It is generally easy to analyse such data using an analytical model f, which can take any value of x as input and spits out a model estimate of the true value directly: $z = f(x, \theta)$. But when *time* is an important model input, we often use dynamic models that cannot be solved analytically. We have mentioned the common example of process-based 'computer' models (PBMs). Such models typically require continuous time series of inputs $\{x(t)\}_{t=1..n}$, which they use to simulate system behaviour iteratively as $z(t + 1) = z(t) + f(z(t), x(t), \theta)$. Such models produce more output than there are data: there are more model input-output combinations $(x(t), z(t))$ than

© Springer Nature Switzerland AG 2020
M. van Oijen, *Bayesian Compendium*,
https://doi.org/10.1007/978-3-030-55897-0_7

there are observations (x, y). So in Bayesian calibration of PBMs we always need to be careful to select the subset of model outputs that corresponds to observations.

Because of these differences between analytical models and PBMs, it is hard to write very general code for Bayesian calibration using the Metropolis algorithm. So let's begin modestly, by writing code that works for most analytical models. We shall consider the case of models with multivariate output, such as PBMs, in Chap. 9. Another problem, especially with R, is that it is hard to write code that works both with scalars and vectors, and with univariate as well as multivariate model inputs and/or outputs. The code examples in this chapter aim for readability rather than computational efficiency.

7.1.2 The Need for Log-Transformations in MCMC

Probability densities for parameter values that are several standard deviations away from the mean will be very close to zero. And in our MCMCs we shall often find ourselves walking through parameter space far from the mean of the distribution (whose location we do not know when we start the MCMC). But computers have limited precision to represent very small numbers: typically numbers smaller than about 10^{-300} are rounded to zero. So to avoid numerical problems we work mostly with log-transformed prior probabilities and log-transformed likelihoods. Let's take the example of the standard normal distribution with mean 0 and standard deviation 1. A standard computer can still represent the probability density $p[\theta] = N[\theta \mid \mu = 0, \sigma^2 = 1]$ when θ is 38 standard deviations away from the mean, but not when it is 39 or more. Densities are then rounded to zero, which makes it impossible to calculate the probability ratios in the MCMC. However, there is no problem when we use $log(p[\theta])$:

```
  dnorm( 38:40, m=0, sd=1 ) ; dnorm( 38:40, m=0, sd=1, log=T)
> [1] 1.097221e-314   0.000000e+00   0.000000e+00
> [1] -722.9189 -761.4189 -800.9189
```

The problem is most severe for parameter-rich models. With 200 independent parameters that are all just three standard deviations from their mean, a computer rounds their joint probability (the product of the 200 densities) to zero, but there are no problems if we calculate the equivalent sum of logarithms instead:

```
  z <- rep( 3, 200 ) ; prod( dnorm(z) ) ; sum ( dnorm(z,log=T) )
> [1] 0
> [1] -1083.788
```

So in this chapter, the Metropolis algorithm for MCMC that we introduced in Chap. 6 will be written slightly differently, but mathematically equivalently. We will replace the calculation of the probability *ratio* with the calculation of a log-probability *difference*. Let's begin with an example where we want to use the Metropolis algorithm to sample from the posterior distribution of a simple linear model.

7.2 Bayesian Calibration of a 2-Parameter Model Using the Metropolis Algorithm

In this section, we shall be calibrating the straight-line model. So there are two parameters: an intercept and a slope. Our data set will be y which we treat as a function of independent variate x. There will be three data points (x_i, y_i) and for each we formulate our measurement uncertainty as a Gaussian with mean zero and variance equal to 3. The covariance matrix for all three measurements is then a diagonal matrix Sy with the value 3 on the diagonal and zero's elsewhere.

```
x  <- c(10,20,30) ; y  <- c(6.09,8.81,10.66)
ny <- length(y)   ; Sy <- diag(3,ny)
```

We write a short R-function f() to represent our model, and we assign a Gaussian prior parameter distribution $N[mb, Sb]$. We start with a very large prior uncertainty by setting the 2×2 prior covariance matrix Sb to have values of 10^4 on the diagonal. We also encode the calculation of log(prior) and log(likelihood) as R-functions, and our final preparation is yet another R-function that calculates the sum of log(prior) and log(likelihood). We call that function logPost() to indicate that it calculates the logarithm of the posterior density up to an unknown additive constant. Here is the R-code for these choices:

```
f  <- function( x, b ) { cbind(1,x)%*%b }
mb <- c(0,0) ; nb <- length(mb) ; Sb <- diag(1e4,nb)
logPrior <- function( b, mb.=mb, Sb.=Sb )
   { dmvnorm( b, mb., Sb., log=T ) }
logLik   <- function( b, y.=y, Sy.=Sy, x.=x, f.=f ) {
   dmvnorm( y., mean=f.(x, b), sigma=Sy., log=T ) }
logPost <- function( b, mb.=mb, Sb.=Sb, y.=y, Sy.=Sy, x.=x, f.=f ) {
   logPrior(b, mb., Sb.) + logLik(b, y., Sy., x., f.) }
```

7.2.1 The Metropolis Algorithm

Now we come to the key step: implementing the MCMC. We use the Metropolis algorithm for which we already showed code in the last chapter. But there will be two differences here. One is that our modified function MetropolisLogPost works with log-transformations throughout. The second is that the function expects as input the specification of the model f.() whose parameters b we are calibrating. The model is used in the calculation of the likelihood. The fact that we made the model specification one of the inputs for MetropolisLogPost makes that function widely applicable: we can use it to calibrate many different models, as we shall see further below. But we first stay with the straight-line model.

```
MetropolisLogPost <- function( f.=f, logp=logPost, b0=mb,
                               SProp=Sb/100, ni=1e4, ... ) {
  bChain <- matrix( NA, nrow=ni, ncol=length(b0) ) ; bChain[1,] <- b0
  for( i in 2:ni ) {
```

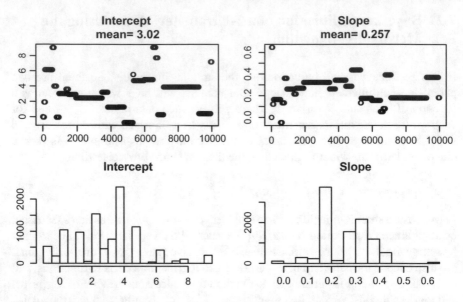

Fig. 7.1 Failed Metropolis sampling of the straight-line model using the default settings for the proposal distribution

```
     b1   <- bChain[i-1,] + as.numeric( rmvnorm(1,sigma=SProp) )
     if( log(runif(1)) < logp(b1,f.=f.,...) - logp(b0,f.=f.,...) )
         bChain[i,] <- b0 <- b1
     else bChain[i,] <- b0 }
return(bChain) }
```

7.2.2   Failed Application of MCMC Using the Default Settings

After all our preparatory work, it has now become very easy to carry out a Bayesian calibration for the two parameters of our straight-line model. If we accept the default settings for chain length and proposal distribution, we can run the calibration with just this code:

```
bChain1 <- MetropolisLogPost( f )
```

So all we need to do now is examine the chain of parameter values bChain1 that we have generated. The results are plotted in Fig. 7.1. But these results are not good at all! The trace plots show that the algorithm got stuck frequently for many iterations without accepting new proposed parameter vectors. In other words, the algorithm mixed poorly so that parameter space was not well explored. And the consequence is that our sample from the posterior does not show smooth, bell-shaped histograms. In MCMC-speak: the algorithm has not converged.

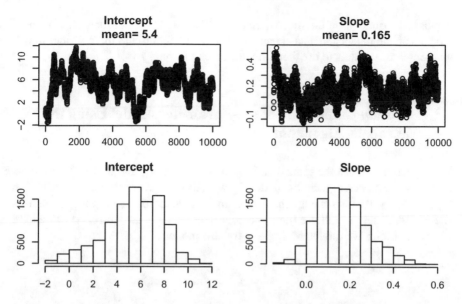

Fig. 7.2 Metropolis sampling of the straight-line model with small proposal variances

So what went wrong? We could say that our chain was not long enough, and the properties of MCMC sampling guarantee that we would get a representative sample from the posterior if we increased the number of iterations manifold. But that is in general not the best response to poorly performing MCMCs. The fact that the vast majority of the 10,000 proposed parameter vectors was rejected means that we should improve our proposal distribution. Remember that we used (our own) default settings for the algorithm, and these included a Gaussian proposal distribution with step length variances equal to 0.01 times the prior variances. But our prior variances were set at the high values of 10^4, so proposal variances were 10^2 and average proposal step lengths were thus about 10. Those step lengths were likely too long to allow the algorithm to properly explore the small part of parameter space that has high posterior probability. So we reduce the proposal step lengths and run the algorithm again:

```
bChain2 <- MetropolisLogPost( f, SProp=diag(c(0.1,0.01)) )
```

Reducing the proposal step lengths makes the algorithm perform much better, as we can see in Fig. 7.2. We now do see convergence in the trace plots and smooth posterior histograms.

Another lesson we can learn from the two MCMCs is the importance of prior information. If we had known beforehand that the intercept would be between 0 and 10 and the slope between 0 and 1, then we could have used smaller proposal steps from the beginning. But what we have done is a common solution: use a first MCMC to test a default proposal distribution, discard its results, and try again with better proposals. There are actually so-called *adaptive MCMC-algorithms* that optimise the proposal distribution on the fly (Andrieu and Thoms 2008), but their effectiveness depends on the model and data at hand, and we shall not discuss them further in this chapter (some pointers to the literature are given in Chap. 10).

7.3 Bayesian Calibration of a 3-Parameter Model Using the Metropolis Algorithm

In this section, we use the same definition of the logPrior() and logLik() functions, but a new 3-parameter quadratic model fq with its own prior parameter distribution $N[mbq, Sbq]$. We also change the data to make a quadratic model more plausible. The new data set yq will use the same covariate values x as before. After making those definitions, we run the Metropolis for this model:

```
x          <- c(10,20,30) ; yq <- y + 0.1*x^2
nyq        <- length(yq)  ; Syq <- diag(30,nyq)
fq         <- function( x, b ) { cbind(1,x,x^2)%*%b }
nbq        <- 3 ; mbq <- rep(0,nbq) ; Sbq <- diag(nbq) * 1e4
bqChain <- MetropolisLogPost( fq, b0=c(0,0,0),
   SProp=diag(c(1,0.1,0.01)), mb.=mbq, Sb.=Sbq, y.=yq )
```

The results of this Bayesian calibration are shown in Fig. 7.3. There are a few more lessons here. If we study the trace plot for the second slope (the regression coefficient for the quadratic term), then the estimates initially seem to be unstable but gradually they reach values around 0.1 and then the algorithm stays around there. Such an initial phase of instability or drift is standard for MCMC. It arises when we start the chain at an arbitrary point in parameter space (often the mean of the prior) that may be far removed from the area with high posterior probability. So the algorithm needs a number of iterations before it finds the primary area to sample from. That initial drifting phase is called the *burn-in* and we should remove that part of the chain from our posterior sample because burn-in points are not representative. Looking again at the trace plot, it seems that the burn-in for the second parameter is only a few thousands iterations but that is actually wrong! The burn-in is defined for the *joint* probability distribution of all parameters. If one or more parameters still show drift, then others that may seem to have stabilised cannot be trusted to stay where they are because parameter estimates are not independent. We must conclude that the MCMC of Fig. 7.3 has not converged yet, so the chain length should be extended.

7.4 More MCMC Diagnostics

We conclude with mentioning a few more analyses that are good to do after you have carried out an MCMC. You have already seen the value of the trace plots for visually assessing whether the MCMC has converged. [There are also semi-formal methods for quantifying the degree of convergence which we shall discuss in a later chapter.]

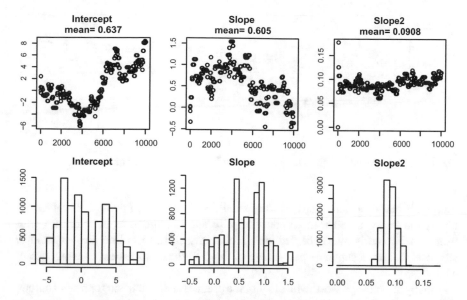

Fig. 7.3 Metropolis sampling of the 3-parameter quadratic model

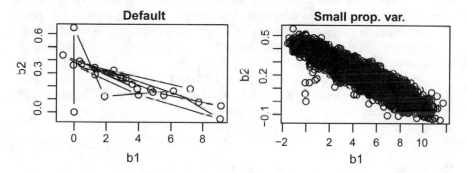

Fig. 7.4 Scatter plots for the two MCMCs of the linear model

But remember that the posterior distribution is a joint distribution, so trace plots for individual parameters do not tell the whole story. You can make scatterplots to study posterior correlations between parameters (Fig. 7.4). Do you recognise the burn-in phase in the panel on the right?

Posterior parameter correlations can of course also be assessed in the form of the posterior covariance or correlation matrix, but those only quantify the strengths of linear relationships, whereas scatter plots can reveal any non-linear interdependencies too.

While you are optimising your proposal distribution, it may be useful to calculate what fraction of the proposals was accepted. But do not aim for excessively high proposal acceptance, because that can be a sign of too small proposal steps and poor

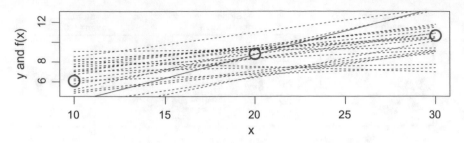

Fig. 7.5 Twenty predictions from the posterior distribution for the linear model

mixing. When proposals steps are very small, every proposed parameter vector has nearly the same value for the product of prior and likelihood as the current vector in the chain, and is thus accepted with probability close to 1. Practical experience suggests that an acceptance percentage of 20–50% leads to fastest convergence of MCMCs.

Finally, it is always worthwhile to plot the predictive distribution for model output. So far, all plots in this chapter have been about the parameters that we are calibrating, but we of course calibrate them to make our models perform better. After running your MCMC, you can take a subsample from the chain (e.g. an equidistant sample like the parameter vectors at iterations 100, 200, .. after burn-in). Figure 7.5 shows a sample from the predictive distribution from our second (successful) MCMC for the linear model.

In the next chapter, we shall be comparing Bayesian methods for the linear model, including calibration by MCMC, to many different methods that you may already be familiar with.

Exercises

1. Identifiability of parameters. Use the same data, likelihood function and prior as before for linear regression, but use a three-parameter model with a clearly redundant formulation: f <- function(x,b){ cbind(1,x,x) %*%b }. Note the double appearance of the covariate x. Carry out the Bayesian calibration for this model using MCMC and check the sample from the posterior distribution. Are the marginal posterior distributions for the two slope-parameters b[2] and b[3] the same? Examine the variances for both slope parameters. Why are both variances much larger than the variance of the single slope-parameter in the regular 2-parameter linear model? What is the uncertainty of the *sum of b[2] and b[3]*? How does that compare to the uncertainty of the slope-parameter in the regular model?

2. Convergence and posterior. But what about the MCMC, do you think it will converge more quickly for the 2-parameter model than for the 3-parameter model? To have the MCMCs perform identically, how should the prior parameter distribution

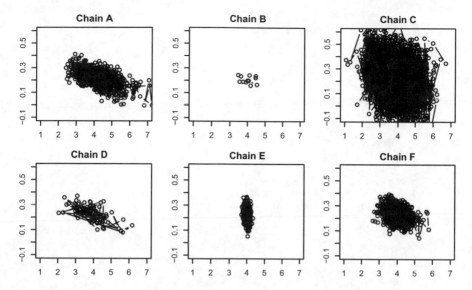

Fig. 7.6 Exercise: Try matching panels A-F to MCMCs 1–6 described in the text

and the proposal distribution for the 3-parameter model be set relative to the settings for the 2-parameter model?

3. Which MCMC produced the following posterior samples? This exercise is about Fig. 7.6. Each panel in the figure shows the posterior sample from a MCMC. Your task is to provide a correct legend for the figure, i.e. say which panel comes from which MCMC. In all cases, we calibrated the straight-line model as we did earlier in this chapter. But we varied data uncertainty, the prior, the proposal distribution and the number of MCMC iterations. For the purposes of this exercise, the standard settings are `mb=c(4,0.2); Sb=diag(1,2); Sy=diag(3,3); SProp=diag(0.1,2); ni=5e3`. The following MCMCs were carried out:

- 1: Standard settings.
- 2: Variance of measurement uncertainty increased by a factor 10.
- 3: Prior variance for the intercept reduced by a factor 100.
- 4: Prior variances for both parameters decreased by a factor 5.
- 5: Proposal variance increased by a factor 50.
- 6: Number of MCMC iterations reduced by a factor 100.

Now try to match figure panels A-F to these MCMCs 1–6. We give one answer already: the first panel shows the results from using the standard settings. So panel A = MCMC 1.

Chapter 8
Twelve Ways to Fit a Straight Line

8.1 Hidden Equivalences

Fitting a straight line through data can be done in many ways that may seem different at first, but after closer inspection prove to be based on the same mathematics. In this chapter, we shall fit a line to data in twelve different ways and compare the resulting parameter estimates. So our goal is to estimate the intercept and the slope of the line. This is our line-up of algorithms:

1. Ordinary Least Squares (OLS) = R-function 'lm'
2. Weighted Least Squares (WLS)
3. Generalised Least Squares (GLS)
4. Bayesian calibration analytically (Lindley and Smith 1972)
5. Kalman Filtering (KF)
6. Conditional multivariate Gaussian
7. Gaussian Network graphical model (GM)
8. Gaussian Process (GP)
9. Accept-Reject Monte Carlo sampling (A-R)
10. Metropolis MCMC
11. Gibbs Sampling MCMC
12. R-package JAGS (also MCMC)

We are going to assume that we have the same measurement uncertainty for each data point, and that our prior uncertainty about the two parameters is very large. It then turns out that the first eight methods will give exactly the same results for the parameter estimates, because they are all based on the same underlying mathematics for Gaussian linear models. The final four algorithms will also give the same results, but with minor deviations because of sampling variability.

The goal of this exercise is threefold. First, I like to show that ostensibly 'classical' methods, such as the least squares estimations, are the same as Bayesian calibration given certain assumptions for the prior and likelihood. Secondly, I hope that the exercise, by convincing you of the strong unity underlying most statistical approaches,

© Springer Nature Switzerland AG 2020
M. van Oijen, *Bayesian Compendium*,
https://doi.org/10.1007/978-3-030-55897-0_8

will reassure you that there is much less to learn than confusing terminology might suggest. Finally, this chapter will be a gentle first introduction to methods that we will study in greater detail in later chapters.

8.2   Our Data

We shall mainly be using the following data in our examples.

```
x <- c(10,20,30) ; y <- c(6.09,8.81,10.66) ; ny <- length(y)
```

The vector x contains 3 values of our scalar predictor variable (which could for example be time or distance or temperature) and the vector y contains the corresponding observations for the dependent variable.

8.3   The Normal Equations for Ordinary Least Squares
Regression (OLS)

Say we want to fit a straight line to our data. Classical statistics has provided us with simple equations that we can use for this purpose, called the Normal Equations:

$$\underline{\text{NORMAL EQUATIONS}} \text{ for } y = \beta_1 + \beta_2 x :$$
$$\text{Estimate of slope: } \hat{\beta}_2 = \frac{Cov(x,y)}{Var(x)}$$
$$\text{Estimate of intercept: } \hat{\beta}_1 = \bar{y} - \hat{\beta}_2 \bar{x} \tag{8.1}$$
$$\text{Correlation coefficient: } \rho = \frac{Cov(x,y)}{\sigma_x \sigma_y},$$

where the hat-symbol $\hat{}$ stands for 'estimated value of' and the overbar-symbol $\bar{}$ for 'mean of'. Using these equations is referred to as Ordinary Least Squares regression, or OLS for short. Let's program the equations in R.

```
b2 <- cov(x,y) / var(x)
b1 <- mean(y) - b2 * mean(x)
```

Running these two lines of code produces the following results.

$$\underline{\text{ORDINARY LEAST SQUARES (OLS):}}$$
$$\hat{\beta}_1 = 3.95; \quad \hat{\beta}_2 = 0.2285. \tag{8.2}$$

We see that the value of the intercept is estimated as 3.95, and the value of the slope as 0.2285. The equations are simple and easy to implement, but we could have

Fig. 8.1 Linear regression
on three data points

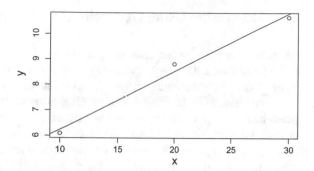

chosen to use the R-function 'lm' which also carries out OLS using the Normal
Equations, and therefore gives the same results:

```
lm.y <- lm(y~x) ; print(coef(lm.y))
> (Intercept)              x
>      3.9500        0.2285
```

Figure 8.1 shows the data and the line that we have fit.

8.3.1 Uncertainty Quantification

OLS is not Bayesian. In Bayesian modelling, we need to state how we treat uncertainty about data error and what priors we assign to parameters. In non-Bayesian
methods such as OLS using R's lm-function, those parts of the problem are swept
under the carpet: data uncertainty is estimated from the residuals when fitting the
model, and priors are assumed to be uniform on the domain from minus to plus
infinity.

So if we use the Normal Equations or 'lm', we do not have to specify our uncertainty about errors in the measurements y. Therefore, without further information,
this method cannot quantify uncertainty about the estimated parameter values. But if
we assume that: (1) our model and parameter estimates are correct, (2) measurement
errors are all drawn independently from the same zero-centered Gaussian distribution, then we can estimate the variance of that error distribution from the properties of
the residuals. And then we can quantify what that measurement uncertainty implies
for uncertainty about the intercept and slope. In fact, with these assumptions there are
simple analytical formulas for the posterior variances of slope and intercept as well as
their covariance. But instead of giving these, we immediately move to more general
and powerful equations that do not require measurement errors to be independent
and identically distributed. The solution for OLS is then just a special case.

8.4   Regression Using Generalised Least Squares (GLS)

We now fit the line to our data using the equations for Generalised Least Squares (GLS) estimation. This requires us to specify uncertainty about measurement errors in the form of a covariance matrix Σ_y, with the variances on the diagonal and the error-covariances in the other positions. GLS allows us to have more than one independent predictor variable, so instead of just x, we could have $x^{(1)}, .., x^{(n_x)}$. We collect the predictor variables in the so-called *design matrix* $\mathbf{X}$, which is a matrix of magnitude $n_y \times (n_x + 1)$ with a column of ones on the left followed by n_x different columns for $x^{(1)}, .., x^{(n_x)}$. We also collect the parameters in one vector $\beta = (intercept, slope(1), .., slope(n_x))$.

The GLS-equations can then be written very compactly as:

$$\text{Model: } y = \mathbf{X}\beta + \epsilon \quad with \, \epsilon \sim N[0, \Sigma_y]$$

$$\text{Estimate of covariance matrix: } \widehat{\Sigma_\beta} = (\mathbf{X}^\top \Sigma_y^{-1} \mathbf{X})^{-1} \tag{8.3}$$

$$\text{Estimate of parameter vector: } \hat{\beta} = \widehat{\Sigma_\beta} \mathbf{X}^\top \Sigma_y^{-1} y.$$

In this chapter, we only have one predictor variable, so $\mathbf{X}$ will be a $n_y \times 2$ matrix, and β will just be (slope, intercept). Let's now implement the GLS-equations in R. As mentioned, that requires us to specify the covariance matrix for measurement error Σ_y, which we initially shall choose to be a diagonal matrix with the same value everywhere on the diagonal:

$$\Sigma_y = \begin{bmatrix} 3 & 0 & 0 \\ 0 & 3 & 0 \\ 0 & 0 & 3 \end{bmatrix} \tag{8.4}$$

We now implement the GLS-equations in R.

```
X    <- cbind( 1, x )                          # Design matrix
Sb_y <- solve(    t(X) %*% solve(Sy) %*% X )   # GLS variance
mb_y <- Sb_y %*% t(X) %*% solve(Sy) %*% y      # GLS estimator
```

When running the code, we get the following results.

$$\text{GENERALIZED LEAST SQUARES (GLS):}$$

$$\hat{\beta} = \begin{bmatrix} 3.95 \\ 0.2285 \end{bmatrix}; \quad \widehat{\Sigma_\beta} = \begin{bmatrix} 7 & -0.3 \\ -0.3 & 0.015 \end{bmatrix}. \tag{8.5}$$

So we see that we get the same estimates for the slope and the intercept as with OLS, but we also get a covariance matrix for the parameters, representing our uncertainty about those estimates.

8.4.1 From GLS to WLS and OLS

To retrieve the OLS-equations from GLS, the covariance matrix for measurement error must have the same value everywhere on the diagonal, and zeros everywhere else. Such a diagonal matrix can be written as a scalar σ_y^2 times a $n_y \times n_y$ identity matrix: $\Sigma_y = \sigma_y^2 \mathbf{I}$. And that is in fact what we did in our example, which therefore was both GLS and OLS.

The example of GLS that we gave here is also an example of Weighted Least Squares (WLS) regression which is intermediate between OLS and GLS. WLS allows the different data points to have different error variances (which will act as inverse weights), but requires that the data uncertainties are not correlated. So the measurement error covariance matrix in WLS must be diagonal, as was the case for our Σ_y. In short:

- OLS: Measurement errors uncorrelated and equal variance $\rightarrow \Sigma_y = \sigma_y^2 \mathbf{I}$,
- WLS: Measurement errors uncorrelated and unequal variance $\rightarrow \Sigma_y = \text{diagonal}$,
- GLS: Measurement errors correlated and unequal variance $\rightarrow \Sigma_y = \text{symmetric}$ positive definite.

Why do we pay so much attention to these classical equations for linear regression in a book about Bayesian methods? Well, the GLS-equations can be interpreted as Bayesian linear regression using Gaussian priors and likelihoods. GLS is the special case where the Gaussian prior has such a large variance that it effectively becomes a uniform prior.

8.5 The Lindley and Smith (LS72) Equations

The GLS can be generalized to a fully Bayesian approach where the prior can be informative, i.e. the prior parameter variances do not need to be infinite. Lindley and Smith (1972) (hereafter LS72) proved this by deriving equations for the posterior mean and covariance matrix of the linear model with Gaussian distributions for parameter prior and measurement error uncertainty. We shall implement their equations in a moment, but let's first specify a zero-mean bivariate Gaussian prior for the parameter vector β.

PRIOR FOR β:

$$p[\beta] = N[\mu_\beta, \Sigma_\beta], \text{ where}$$

$$\mu_\beta = \begin{bmatrix} 0 \\ 0 \end{bmatrix}; \quad \Sigma_\beta = \begin{bmatrix} 10000 & 0 \\ 0 & 10000 \end{bmatrix}.$$

(8.6)

And here is the very short code for the LS72 equations:

```
Sb_y_equiv <- solve( solve(Sb) + t(X) %*% solve(Sy) %*% X ) ;
mb_y_equiv <- Sb_y_equiv %*% (solve(Sb) %*% mb + t(X) %*% solve(Sy) %*% y)
```

And when you run that code, you get the same results as with GLS (Eq. (8.5). The results are the same despite the introduction of the prior probability distribution for the two model parameters in Eq. (8.6) because we set the prior variances so high that only the likelihood was informative, which is the underlying assumption of GLS. By the way, you can see that the GLS equations are a special case of the LS72 equations with the prior parameter variance going to infinity and Σ_β^{-1} thus going to zero. So if we remove the terms involving Σ_β^{-1} from the LS72 equations (i.e. `solve(Sb)` in our R-code), we retrieve the GLS equations that we showed above.

8.6   Regression Using the Kalman Filter

Now let's look at equations that at first look completely different from the equations shown above, but are actually mathematically the same: the Kalman Filter equations (Soyer 2018).

The Kalman Filter (KF) is commonly used as a procedure for updating our estimate for the state of a dynamical system, each time when new data come in (and that is how we introduce it in Chap. 21). The KF assumes that (1) state-change follows a linear function of the state itself, with known parameters, and (2) state uncertainty can be represented by a Gaussian distribution. It is highly unusual, and really not very practical, to use the KF for linear regression, i.e. for parameter estimation rather than for state estimation. But we shall do so anyway to show that, under the hood, the KF uses the same mathematics as the GLS-equations.

The trick to using KF for linear regression is to realise that there is no real difference between the state of a system at a specific time, and a parameter vector. Both are uncertain quantities that we want to learn about. The fact that the state of a dynamical system can be written as a function of the state that it had in the recent past, does not alter that fact - it just helps us constrain state uncertainty. So the use of KF for linear regression requires us to change our perspective: we treat the parameter vector β of the linear model (the slope and intercept) as the 'state' to be estimated with KF, and the sequence of values for the independent variable x as the known 'parameter' vector. The data y will as usual provide the information we need to carry out the estimation.

KF is a fully Bayesian procedure. As for GLS, KF requires us to begin with specifying Gaussian probability distributions for measurement errors, but it adds to that the requirement to specify a Gaussian prior for what we want to estimate (normally in KF the system state but here the parameter vector β), and we use the same prior that we defined above for the Lindley & Smith equations.

So at this stage we have the data y with their uncertainties and the prior $p[\beta]$. The final ingredient we need for KF is the linear model itself, which in KF-parlance is referred to as the *observational operator matrix*, and is typically denoted as **H**. But this matrix is in fact exactly the same as the design matrix that we already defined above, so we denote it as $\mathbf{X}$. The idea behind the KF-terminology is that the product of the observational operator with the state-vector plus measurement error predicts the observations y. So here $\mathbf{X}$ is again a 3×2 matrix with all ones in the first column and the values of x in column 2. That completes the specification of our linear model, and KF-theory now tells us that the solution for $p[\beta|y]$ is as follows:

$$\text{'Kalman Gain':} \quad \mathbf{K} = \Sigma_\beta \mathbf{X}^\top (\mathbf{X} \Sigma_\beta \mathbf{X}^\top + \Sigma_y)^{-1}$$

$$\text{Posterior mean: } \mu_{\beta|y} = \mu_\beta + \mathbf{K}(y - \mathbf{X}\mu_\beta) \tag{8.7}$$

$$\text{Posterior covariance matrix: } \Sigma_{\beta|y} = (\mathbf{I} - \mathbf{KX})\Sigma_\beta$$

where $\mathbf{X}^\top$ is the transpose of $\mathbf{X}$ (rows and columns interchanged) and $\mathbf{I}$ is the identity matrix (ones on the diagonal, zeros everywhere else). We implement this in R:

```
KalmanGain <- function(B,H,R){ B %*% t(H) %*% solve(H %*% B %*% t(H) + R) }
K          <- KalmanGain( Sb, X, Sy )
mb_y       <- mb + K %*% ( y - X %*% mb )
Sb_y       <- ( diag(nb) - K %*% X ) %*% Sb
```

When you run this code, you find that the estimates for β that we find with KF are the same as with GLS and LS72. The fact that KF for linear regression gives the same parameter estimates and posterior covariance matrix shows that the methods are mathematically identical—at least if the parameter prior that we specify in KF is uninformative, which was the choice we made by assigning very large variances in our Σ_β.

8.7 Regression Using the Conditional Multivariate Gaussian

In the fully Gaussian world, Bayesian estimation is just a matter of manipulating mean vectors and covariance matrices. Let $z = (\beta, y)$ represent the combination of parameters and observations (that we have not seen yet). We represent our prior uncertainty about the parameters and the data in the Gaussian distribution with mean vector μ_z and covariance matrix Σ_z. What will be our posterior distribution for β once we get observational values for y?

We shall now define an R-function that calculates the mean and covariance matrix of the posterior conditional multivariate Gaussian distribution $p[\beta|y]$. The function takes as arguments μ_z, Σ_z and a vector with observed true values y and produces as output the list $(\mu_{\beta|y}, \Sigma_{\beta|y})$.

```
GaussCond <- function( mz, Sz, y ) {
  i <- 1 : ( length(mz) - length(y) )
  m  <- mz[i]   + Sz[i,-i] %*% solve(Sz[-i,-i]) %*% (y-mz[-i])
```

```
    S   <- Sz[i,i] - Sz[i,-i] %*% solve(Sz[-i,-i]) %*% Sz[-i,i]
    return( list( m=m, S=S ) ) }
```

Now we use this function for linear regression. But how do we determine
the covariance matrix for parameters and unobserved data? In this case of linear regression with just three data points, it is not too much work to create this
5×5 matrix. The top left 2×2 will just be the prior covariance matrix for the
parameters Σ_β. So the first two entries on the diagonal are the prior variances
for the parameters which we denote as $V_{\beta[1]}$ and $V_{\beta[2]}$. The three other entries on
the diagonal are the prior *unconditional* variances for the as yet unobserved data
$V_{y[i]}^{unc}$, i=1:3. These variances will be large because they are not conditioned on β.
Under our linear model, the unconditional observation variances are calculated as
$V_{y[i]}^{unc} = V_{\beta[1]} + x_i^2 V_{\beta[2]} + V_{y[i]}$. Covariances between observations are calculated as
$Cov_{y[i],y[j]}^{unc} = V_{\beta[1]} + x_i x_j V_{\beta[2]}$. That leaves the covariances between parameters and
observations, which we find as $Cov_{\beta_1,y[i]}^{unc} = V_{\beta[1]}$ and $Cov_{\beta_2,y[i]}^{unc} = x_i V_{\beta[2]}$. This may
all look complicated but it can be implemented in the following two lines of R-code.

```
    mz <- c( mb, X %*% mb )
    Sz <- rbind( cbind( Sb          ,    Vb*t(X)                ),
                 cbind( t(Vb*t(X)), t(Vb*t(X))%*%t(X)+Sy ) )
```

The prior mean and covariance matrix are then found to be:

PRIOR MULTIVARIATE GAUSSIAN OF PARAMETERS AND OBSERVATIONS

$$
\mu_z = \begin{bmatrix} 0 \\ 0 \\ 0 \\ 0 \\ 0 \end{bmatrix} ; \quad \Sigma_z = \begin{bmatrix} 10000 & 0 & 10000 & 10000 & 10000 \\ 0 & 10000 & 1e+05 & 2e+05 & 3e+05 \\ 10000 & 1e+05 & 1010003 & 2010000 & 3010000 \\ 10000 & 2e+05 & 2010000 & 4010003 & 6010000 \\ 10000 & 3e+05 & 3010000 & 6010000 & 9010003 \end{bmatrix} .
\tag{8.8}
$$

All we have to do now to update this prior multivariate Gaussian—and find the
Bayesian posterior—is condition the distribution on the observations. We do this by
means of our previously defined R-function GaussCond.

```
    Post <- GaussCond( mz, Sz, y ) ; mb_y <- Post$m ; Sb_y <- Post$S
```

And, this gives the same results as GLS, LS72 and KF, as you can verify for
yourself.

8.8   Regression Using Graphical Modelling (GM)

The complicated derivation of the joint covariance matrix for parameters and observations of the preceding section can be much simplified by the use of Graphical
Modelling. This technique will be introduced more fully in Chap. 15. But the basic

Fig. 8.2 Directed Acyclic Graph (DAG) for the linear regression graphical model

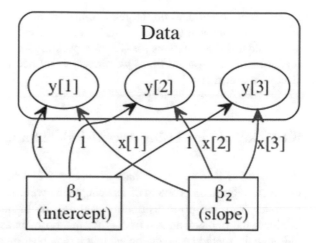

idea is simple: every GM represents a joint probability distribution as a graph with arrows between nodes that show how the joint distribution can best be decomposed into a number of simpler conditional probability distributions. In the present case of our regression problem, we define a network with five nodes: the two parameters of the straight line model and the three observations, and we show the graph for the network in Fig. 8.2.

The theory of GM has provided us with elegant algorithms for deriving the desired covariance matrix from that graph structure (Shachter and Kenley 1989). We shall show those algorithms, including R-code, in Chap. 15, but for now just state that the GM-approach leads to the exact same covariance matrix as we cumbersomely derived in the preceding section. It therefore also leads to the same posterior distribution for the parameters in our regression problem as we have seen many times before in this chapter.

8.9 Regression Using a Gaussian Process (GP)

We shall encounter Gaussian Processes (GP) at various places in this book, with the most detailed description in Chap. 14. GPs are not just multivariate but 'infinite-variate' Gaussian distributions that come with a deterministic mean function for trend modelling plus a covariance function for correlations between any two of the infinite variates. That short, cryptic description will make more sense after having read the first part of Chap. 14, but for now it may be enough to state how a GP can be used for the regression problem of this chapter. If the correlation length ϕ for a GP is very small, close to zero, then even points that are close to each other behave independently. In other words, the trend-part of the GP dominates and the role of the stochastic process component becomes negligible apart from quantifying the variance for data error uncertainty. Because the data points are then effectively independent conditional on the regression parameters, this approach of "GP with $\phi \rightarrow 0$" is equivalent to linear regression if the GP-trend is modelled as linear. So you could use the equations for GP (or a package for GP such as geoR) for linear regression if you wanted to, as we already described in the chapter on emulation. [But why would you?]

8.10   Regression Using Accept-Reject Sampling

We showed A-R in action in Chap. 7 (see Fig. 5.1), and it can of course be used for the linear regression problem of this chapter as well. All we need to do is generate a sample from our prior distribution and use the likelihood function for pruning. We leave the implementation of A-R for this problem as an exercise for the reader, although the method is so inefficient that it may be difficult to get it to work without needing an unfeasibly large sample. Rather than sampling from the full unbounded prior parameter space that we have here, you may want to limit yourself to a small part of parameter space from the beginning.

8.11   Regression Using MCMC with the Metropolis Algorithm

We already showed linear regression by means of MCMC with the Metropolis algorithm in Chap. 7. You can get very close to the results of all the methods shown above by using a chain length of 10^5.

8.12   Regression Using MCMC with Gibbs Sampling

We shall now fit the straight line by means of another MCMC algorithm, *Gibbs sampling*. But before we show how this can be implemented, we define an R-function GaussMult that will come in handy when writing code for Gibbs. GaussMult calculates the mean and variance of a Gaussian that is formed by multiplying (and then renormalising) n different univariate Gaussians, each with their own mean and variance.

```
GaussMult <- function( m=rep(0,2), V=rep(1,2) ) {
    n        <- length(m)
    weights <- sapply( 1:n, function(i){ prod(V[-i]) } )
    m_Mult  <- sum( m * weights ) / sum( weights )
```

```
V_Mult  <- prod( V )                    / sum( weights )
return( list( m=m_Mult, V=V_Mult ) ) }
```

In the Gibbs sampling, we shall be using that function to calculate the product of the marginal prior and likelihoods (one for each data point) for individual parameters. We shall also be using diferent ways of writing the likelihood function than we are used to so far. You can verify yourself that the three lines of R-code for L below give exactly the same value.

```
i  <- 1 ; sy=sqrt(Vy) ; b1 <- 0 ; b2 <- 0
L <- dnorm( b1 + b2*x[i] - y[i], mean= 0              , sd=sy[i]       )
L <- dnorm(           b2         , mean=(y[i] - b1)/x[i], sd=sy[i]/x[i] ) / x[i]
L <- dnorm( b1                   , mean= y[i] - b2 *x[i], sd=sy[i]       )
```

We shall use the last two lines in the R-code for finding analytical solutions for $p[b_1|b_2, y]$ and $p[b_2|b_1, y]$, i.e. the conditional posterior probability distributions for one parameter given all data and the other parameter. That is the basis of Gibbs sampling: taking each parameter one at a time. Our whole Gibbs-algorithm is the following:

```
ni <- 1e3 ; bChain <- matrix(NA, nrow=ni, ncol=2)
b1 <- mb[1] ; b2 <- mb[2] ;  bChain[1,] <- c( b1, b2 )
for (i in (2 : ni)) {
  mb1_L <- y - b2*x ; Vb1_L <- diag( Sy )
  Post1 <- GaussMult( m=c(mb[1],mb1_L), V=c(Vb[1],Vb1_L) )
  mb1   <- Post1$m ; Vb1 <- Post1$V
  b1    <- rnorm( 1, mean=mb1, sd=sqrt(Vb1) )
  mb2_L <- ( y - b1 ) / x ; Vb2_L <- diag( Sy/(x^2) )
  Post2 <- GaussMult( m=c(mb[2],mb2_L), V=c(Vb[2],Vb2_L) )
  mb2   <- Post2$m ; Vb2 <- Post2$V
  b2    <- rnorm( 1, mean=mb2, sd=sqrt(Vb2) )
  bChain[i,] <- c( b1, b2 ) }
```

Gibbs sampling cannot always be used: it requires that we are able to write down the formula for the conditional posterior probability for one parameter given the values of all other parameters and the data. In linear regression, Gibbs can be used because we know how to write down those equations as you can see in the R-code. Another downside of Gibbs is that it considers one parameter at a time, and not the whole parameter vector as in our Metropolis implementation. So you might expect Gibbs sampling to take longer than Metropolis. But that is often not the case because Gibbs has the major advantage that every proposal is accepted. For our example in this chapter, Gibbs is better than Metropolis. The following results were produced with Gibbs using a chain length of just 1000:

GIBBS SAMPLING:

$$\mu_{\beta|y} = \begin{bmatrix} 4.046 \\ 0.225 \end{bmatrix}; \quad \Sigma_{\beta|y} = \begin{bmatrix} 7.319 & -0.309 \\ -0.309 & 0.015 \end{bmatrix}. \tag{8.9}$$

Figure 8.3 shows the frequency distributions for the two parameters generated by the Gibbs sampling.

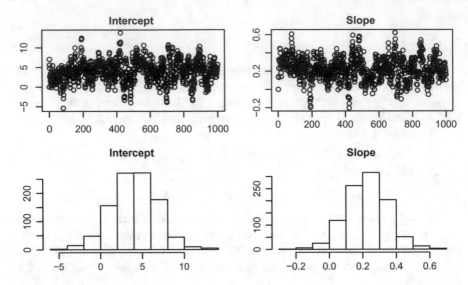

Fig. 8.3 Gibbs sampling using own R-code

8.13   Regression Using JAGS

Finally, and exceptionally for this book, we shall be using the JAGS-software to do the Gibbs sampling for us. The downside of JAGS, and similar software, is that it does not show its actual sampling algorithm, so you cannot improve that where necessary. However, for statistical models that can be expressed in the JAGS language (which excludes, for example, process-based models), JAGS often works very well. The JAGS-code below, which can be run from R if the package rjags has been installed, is largely self-explanatory, but you can find more information in Chap. 16. It is also very fast and produces for our problem the results shown in Fig. 8.4.

```
f.JAGS <- " model { for (i in 1:n) {
    y[i]        ~ dnorm( y.hat[i], tau.y[i] )
    y.hat[i] <- b0 + b1 * x[i]
    tau.y[i] <- pow( sy[i], -1 ) }
  b0 ~ dnorm( 0, 1e-4 ) ; b1 ~ dnorm( 0, 1e-4 ) } "
writeLines( f.JAGS, con="f.JAGS.txt" )
f.JAGS.data <- list ( n=ny, y=y, x=x, sy=diag(Sy) )
f.JAGS       <- jags.model( "f.JAGS.txt", data=f.JAGS.data, n.adapt=5e3 )
update( f.JAGS, n.iter=5e3 )
f.JAGS.codaSamples <- coda.samples( f.JAGS, var=c("b0","b1"), n.iter=5e3 )
f.JAGS.mcmcChain    <- as.matrix( f.JAGS.codaSamples )
```

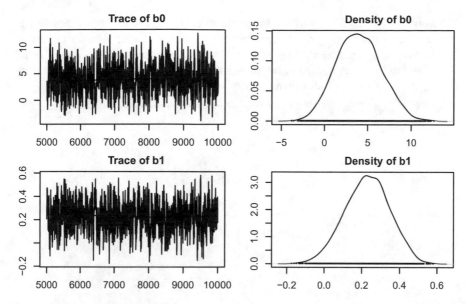

Fig. 8.4 Gibbs sampling using JAGS

8.14 Comparison of Methods

So why do all the above methods give the same results? Well, the primary answer is that they are all correct ways of calculating the Bayesian posterior for the simple case of linear Gaussian modelling. Moreover, the eight analytical methods all effectively carry out regression using the same formula for a conditional Gaussian distribution. However, there are computational differences, in particular because of the different matrix inversions that the methods carry out.

The numerical, sampling-based methods are less accurate and precise for the given problem, but they have the major advantage of being applicable to far more problems, and we can always improve them by using longer chains, better proposal algorithms or different MCMC algorithms. For the calibration of complex models, such as process-based models, we have no choice but to use numerical sampling algorithms, unless we are satisfied with coarse approximations to the posterior distribution.

Exercises

1. Prior uncertainty. Reduce the prior variances, which we specified as being 10^4 in Eq. (8.6), to more realistic values such as 10^2. Does that increase or decrease the posterior estimates for β? Why?

2. Kalman Filtering (KF) without intercept uncertainty. In our application of KF, we specified a very large prior uncertainty for β. Repeat the KF exactly as before, but with $\Sigma_\beta[1, 1]$ set to zero. How does that change the posterior distribution for β? What kind of linear regression have you just done? Why is the posterior uncertainty for the slope-parameter lower than before?

Chapter 9
MCMC and Complex Models

In this chapter we focus on models with multivariate output. That includes most process-based models (PBMs). Models with multivariate output are not fundamentally different from the simpler models we studied in the previous chapters, we can still write them as functions f of their input consisting of covariates x and parameters θ. But the output $z = f(x, \theta)$ from the models will be multivariate, e.g. time series of different properties of an ecosystem. That does not affect the principles of Bayesian calibration in any way, but it does bring some minor practical complications. In this chapter, we illustrate these issues with a quite simple PBM that as output produces two time series: the development over time of the biomass and leaf area of vegetation.

9.1 Process-Based Models (PBMs)

Process-based models (PBMs) are multivariate functions. They are dynamic models that can usually not be analytically solved, so to calculate their output we would have to use numerical integration, iterating over time steps. The input required by PBMs typically consists of time series of driver variables (the covariates x), and parameters θ which include the initial values for the *state variables*. The iterative process of solving a PBM starts from the initial values of the state variables, and then alternates between calculating rates of change and updating the corresponding state variables.

9.1.1 A Simple PBM for Vegetation Growth: The Expolinear Model

The model that we use in this chapter is the 5-parameter expolinear model, EXPOL5, which was developed and studied by Goudriaan and Monteith (1990). It can be written as a single differential equation:

© Springer Nature Switzerland AG 2020
M. van Oijen, *Bayesian Compendium*,
https://doi.org/10.1007/978-3-030-55897-0_9

$$\frac{dW}{dt} = LUE \cdot I_0 \left(1 - e^{-K \cdot LAR \cdot W}\right); \quad W(0) = W_0, \tag{9.1}$$

where W is biomass (gm^{-2}), LUE is the Light-Use Efficiency ($g\,MJ^{-1}$), I_0 is light intensity ($MJ\,m^{-2}d^{-1}$), K is the light-extinction coefficient ($m^2\,m^{-2}$), LAR is the Leaf Area Ratio ($m^2\,g^{-1}$), and W_0 is the biomass at $t = 0$. Despite its brevity, this equation represents considerable knowledge about vegetation growth. The product of LAR and W is the so-called leaf-area index (LAI) of the vegetation, which is the primary variable that determines resource-use. The exponential term represents *Beer's Law* which quantifies the fraction of light that is transmitted through a medium, so the full bracketed term represents the fractional light interception by the canopy. Multiplied with the incident light intensity, I_0, we arrive at the amount of light intercepted by the vegetation per unit of time and ground area. Finally, the multiplication with the LUE represents the process of photosynthesis corrected for respiration, giving us the vegetation growth rate $\frac{dW}{dt}$. The equation has shown to be suitable to describe the growth from seedling to maturity of many different plant species growing in monoculture.

Remarkably, this sophisticated one-equation process-based model allows for an analytical solution if the light intensity remains constant over the period of interest:

EXPOLINEAR EQUATION:

$$W(t) = \frac{ln[1 + exp(I_0 \cdot K \cdot LAR \cdot LUE \cdot t)\,(exp(K \cdot LAR \cdot W_0) - 1)]}{K \cdot LAR} \tag{9.2}$$

We now want to write an R-function for the expolinear equation, to facilitate later MCMC. We shall be having data for both biomass W and leaf-area index LAI, so we make those two variates the outputs from our function EXPOL5:

```
EXPOL5 <- function( t=0, b=c(10,1,0.007,2,1) ) {
   I0   <- b[1] ; K <- b[2] ; LAR <- b[3] ; LUE <- b[4] ; W0 <- b[5]
   W    <- log( 1 + exp(I0*K*LAR*LUE*t) * (exp(K*LAR*W0)-1) ) / (K*LAR)
   LAI  <- W * LAR
   return( list( "W"=W, "LAI"=LAI ) ) }
```

The "5" at the end of the R-function name 'EXPOL5' is to emphasize that this is a 5-parameter model, with the parameter vector representing ($I0, K, LAR, LUE, W0$). Our prior for the parameters of EXPOL5 shall be a multivariate Gaussian with mean parameter vector mb and covariance matrix Sb:

```
bmin5  <- c(9.9,0.5,0.005,1,0.5)        ; bmax5 <- c(10.1,1.5,0.009,3,1.5)
range5 <- bmax5 - bmin5                 ; var5  <- (range5^2) / 12
mb     <- rowMeans( cbind(bmin5,bmax5) ) ; Sb    <- diag( var5 )
```

9.2 Bayesian Calibration of the Expolinear Model

We want to demonstrate a Bayesian calibration of the parameters of the expolinear model, and here are the (virtual) data that we shall be using:

```
t        <- c(20,60,100) ; y <- c(10,500,1200) ; sd < c(5,15,120)
data_W   <- cbind(t,y,sd)
t        <- c(20,80,100) ; y <- c(0.2,5,4)      ; sd <- c(0.1,1,1)
data_LAI <- cbind(t,y,sd)
data     <- list( "W"=data_W, "LAI"=data_LAI )
```

We shall be using the same function for the logarithm of the prior as in Chap. 7, but we need to define a new R-function for the log-likelihood that can handle the kind of output-list for multiple variates that EXPOL5 produces:

```
logLikList <- function( b, dataList, f.=f ) {
  sum( sapply( names(dataList), function(v){ sum(
    dnorm( f.( dataList[[v]][,1], b ) [[v]],
           mean=dataList[[v]][,2], sd=dataList[[v]][,3], log=T ) ) } ) ) }
```

Because we are using a different likelihood function (logLikList()) than in the previous chapter (where we used logLik()), we also need a slightly adapted version of the function for the sum of log(prior) and log(likelihood):

```
logPostList <- function( b, mb.=mb, Sb.=Sb, dataList, f.=f ) {
  logPrior(b, mb., Sb.) + logLikList(b, dataList, f.) }
```

Now we have everything in place for Bayesian calibration of our process-based model, EXPOL5. We run the exact same R-function MetropolisLogPost that we defined in Chapter 7:

```
bChain <- MetropolisLogPost( f=EXPOL5, logp=logPostList, dataList=data )
```

The five parameter trace plots and parameter histograms are shown in Fig. 9.1.

Note that we included calibration for an environmental driver, namely the radiation parameter $I0$. That is perfectly fine: environmental drivers are just one class of parameters or parameter vectors and can be treated in the same way as the other model parameters. This means that from a Bayesian perspective there is never a need for ad-hoc driver gap-filling algorithms. Gap-filling is a parameter-estimation problem like any other and can be addressed with the same Bayesian calibration approach. Also note that we calibrated the initial value $W0$ of state variable W which is a parameter too.

Finally, we make plots of EXPOL5 outputs vs. data to verify that the posterior mean parameter vector makes sense (Fig. 9.2).

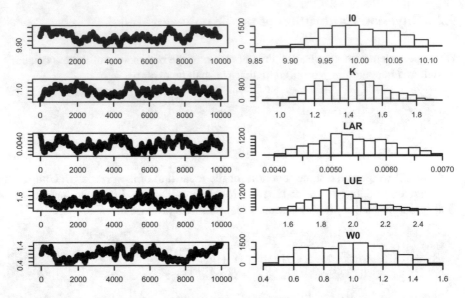

Fig. 9.1 Bayesian calibration of the 5-parameter expolinear model by means of Metropolis sampling

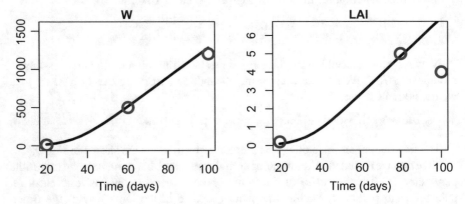

Fig. 9.2 Time series of biomass (W, $g\,m^{-2}$) and leaf-area index (LAI, $m^2\,m^{-2}$), generated by the expolinear equation using the posterior mean parameter vector

9.3   More Complex Models

This chapter showed that the Metropolis algorithm, and MCMC algorithms in general, really are Swiss army knives that can be used for Bayesian calibration of any data set and model, irrespective of the number of variates involved. Bayesian calibration of complex models is gradually becoming more common (see e.g. the review by Van Oijen 2017). If you want to explore such models: some can be downloaded from web repositories. The following three examples of downloadable process-based vegetation models come with code for Bayesian calibration, in each case by means of the Metropolis algorithm:

- BASGRA, a model for grasslands (Van Oijen et al. 2015),
- BASFOR, a model for forests (Van Oijen and Cameron 2020),
- CAF2014, a model for coffee agroforestry systems (Van Oijen 2020).

The following chapter will discuss a variety of issues surrounding the use of MCMC that we have partly skipped over so far.

Chapter 10
Bayesian Calibration and MCMC: Frequently Asked Questions

This chapter answers a number of common questions about Bayesian calibration in general and MCMC in particular. Many topics are addressed elsewhere in this book at greater length (and pointers to the chapters are then given), but some are only addressed briefly here.

10.1 The MCMC Algorithm

1. Can the use of more than one MCMC chain help assess convergence? Yes. If you run more than one chain, each starting from a different point in parameter space, it is much easier to see when the chains have converged. Convergence is apparent when the different chains have moved to the same part of parameter space—and we can establish this formally by testing whether the interchain variance has come down to the level of intrachain variance using, for example, the Gelman-Rubin statistic (Gelman et al. 2013). However, foolproof convergence diagnostics do not exist and visual inspection of the trace plots of a single-chain MCMC remains important.

2. My model is a process-based model: does that affect how I should carry out Bayesian calibration? In principle, not. Bayesian thinking is completely generic and applies to all models and data sets. However, MCMC requires many runs of your model, so if your model is computationally demanding, consider replacing it with a fast emulator (see Chap. 14).

3. My model is perhaps not very good: does that affect how I should run the MCMC? No. The calibration by MCMC takes your model as given. The posterior distribution for the parameters is always conditional on both the data and the model that are being used. We could express this by writing $p[\theta|y, f]$ for our posterior distribution, rather than just $p[\theta|y]$. However, we normally write the shorter version and treat the conditionality on the model as understood. Note that the use of a poor model together with informative data may lead to a posterior distribution for the parameters that will seem unrealistic if considered in isolation—even though conditional on the

© Springer Nature Switzerland AG 2020
M. van Oijen, *Bayesian Compendium*,
https://doi.org/10.1007/978-3-030-55897-0_10

model that posterior is the correct one. The Bayesian calibration will thus help us identify models of poor quality—and in such cases it is advisable to try different models and carry out a Bayesian Model Comparison (BMC; see Chap. 12). There are methods for Bayesian calibration which do not assume that the original model is correct. Instead they add a stochastic term to the model, called the 'discrepancy', to represent the mismatch between model behaviour and reality. Often the discrepancy is modelled as a multivariate Gaussian that quantifies the mismatch for each data point. In short, these methods replace the assumption of the original deterministic model being correct with that of the extended (deterministic + stochastic) model being correct. A problem with that approach is that the stochastic discrepancy term is only defined well near data points and does not help us much with predictions. We discuss the discrepancy at greater length in Chap. 13.

4. Which MCMC algorithm should I use? Metropolis and Gibbs Sampling (Chap. 8) are completely general: they will give the right answer for any combination of model and data—provided the sample from the posterior is large enough. However, there is a multitude of different MCMC algorithms, and new ones keep being developed. Many are adaptive MCMC algorithms in which the proposal distribution changes during the random walk (Andrieu and Thoms 2008). The adaptation method must be carefully designed to guarantee that the algorithm converges to the correct posterior distribution (Robert and Casella 2010). Having said that, the future is likely to see increasing use of adaptive algorithms. Methods that have shown to be useful include DRAM (Haario et al. 2006) and DEMCzs (ter Braak and Vrugt 2008). Among the algorithms for which examples are included in this book, the Metropolis is the one that works best for parameter-rich models such as process-based models (PBM), say more than 10 parameters to be calibrated. Gibbs sampling can generally not be implemented for PBMs because the conditional probability distributions for individual parameters (or groups of parameters) can not be defined. A-R should not be used with parameter-rich models, unless a pre-screening step is added in which the number of to-be-calibrated parameters is reduced to a small number first. A-R has problems with parameter-rich models because it begins by sampling from the prior across the whole of parameter space. This cannot be done exhaustively if the space is high-dimensional. In contrast, MCMC samples directly from the posterior distribution and is efficient in that it spends most time in the area of high probability. However, A-R and related non-MCMC methods such as Importance Sampling have advantages too: they are simpler and easier to apply because there is no proposal distribution to optimise.

10.2   Data and Likelihood Function

See also Chap. 4.

1. Does it matter whether we use all data at the same time or in steps? No. Assume you have two datasets y_1 and y_2. You could use all the data at once to calculate the posterior $p[\theta|y_1, y_2]$. Alternatively, you could first calculate $p[\theta|y_1]$ and then use that as the prior for a Bayesian calibration on y_2. In both cases you arrive at the same final posterior distribution $p[\theta|y_1, y_2]$. This is formally shown in the Appendix on Probability theory (Appendix C) under the heading *Sequential Bayesian updating* and it shows one of the major strengths of Bayesian calibration: it is firmly based on probability theory and always gives the same answer if the same information (in this case y_1 plus y_2) is used.

2. How do I handle systematic measurement error? Systematic measurement error, where all measurements are off by a constant but unknown amount or proportion, can be handled like any other unknown constant: as a parameter that we calibrate. This only tends to be necessary for some types of measurements, e.g. poorly replicated ones. An example may be the measurement of soil emissions of N_2O, where we are always in danger of unrepresentative placement of soil chambers. For such emission data we could use a multiplicative systematic error parameter (for an example of this approach, see Van Oijen et al. 2011).

3. I have a lot of data: do I need to use them all? In principle, yes, as long as the uncertainty about random as well as systematic measurement errors is properly quantified in the likelihood. If the model is very good, then every bit of empirical information will tend to give us a posterior distribution for the parameters which will improve model performance for any application. However, with our imperfect models it may be prudent to aim for balance between different kinds of measurement information. We can do this by not collecting many more data for one variable, one site or one time period, than for others.

4. What do I do if measurement errors are not independent? This is important for calculating the likelihood. We often assume that measurement errors are independent and calculate the overall likelihood of a dataset $p[y|\theta]$ as the product of likelihoods for the individual data points $p[y(i)|\theta]$. This is usually done as a straightforward multiplication of Gaussian probability densities, with the standard deviations σ_i representing each data point's uncertainty. If the measurement errors are correlated we can take that into account by calculating the likelihood differently, e.g. as a multivariate normal distribution where the off-diagonal elements of the variance-covariance are not all zero (Chap. 4). For technical details on this and other methods for modifying the likelihood function, see for example Sivia and Skilling (2006).

5. What do I do if my measurements have uncertainties in the x- as well the y-direction? This can happen if it is unclear at what exact time the measurements were taken. This is easily incorporated in the Bayesian calibration, by adjusting the calculation of the likelihood. Say there is 50% chance that a certain data point $y(t)$ was indeed measured at the reported day and 25% chance each for the days before and after. Then the likelihood for that point is calculated as the weighted average of the likelihoods calculated for each of the three days, with weights 0.25, 0.5 and 0.25.

10.3   Parameters and Prior

See also Chap. 3.

1. Can Bayesian calibration handle correlations between parameters? Yes. Correlations appear whenever we do not know the exact values of two parameters, but do know something about their combined values. For example, we may know that if parameter θ_1 has a high value, parameter θ_2 is likely to have a low value. In fact, each time we speak of "the" parameter distribution, we speak of one single probability distribution for all possible combinations of parameter values, and that distribution typically includes correlations. In other words, the parameters' distribution is a *joint* probability distribution, not a collection of independent distributions for individual parameters. Furthermore, even if the prior distribution has no correlations between parameters, the posterior distribution almost always will.

2. Should I include all my model's parameters in the Bayesian calibration or just a subset? Preferably all of them. Using a subset of the parameters will inevitably lead to underestimation of the contribution of the parameters to model predictive uncertainty. MCMC can handle high-dimensional parameter vectors, but the A-R method will struggle. However, if it is possible to identify a subset of parameters that together account for most of the model behaviour (a practice called 'parameter screening'), then there is little harm in using the subset only.

3. What distribution should I use for the prior distribution? The prior distribution should represent what we know about our model's parameters. That means representing what parameter values are considered impossible, which ones are possible but not likely, which ones are quite plausible etc. If we have no idea about parameter correlations, we can use the product of different distributions for each individual parameter. Using the uniform distribution is generally not a good idea: we typically do not consider all values between some lower and upper limits equally plausible. A triangular distribution is better but still has strange discontinuities. Good choices are the (truncated) normal distribution, the lognormal distribution and the beta distribution. However, there is no golden rule: it all depends on what we can say about our parameter values. Note that there may exist all kinds of complicated constraints on the parameter values that we want to represent in the prior. For example, we may have two parameters θ_{min} and θ_{max} that are both highly uncertain but we know that θ_{min} has to be smaller than θ_{max}. The easiest way to include such information in the prior distribution for the parameters is to reparameterise, e.g. by replacing θ_{min} with a new parameter k_{min}, defined as $\theta_{min}/\theta_{max}$. Reparameterisation may in some cases lead to parameter vectors that are easier to calibrate (Gilks et al. 1995).

10.4   Code Efficiency and Computational Issues

1. How do I make the proposal algorithm of my MCMC method more efficient? Very often, if acceptance rate is low ($<<20\%$), the average proposal step length is too high and vice versa. It may also be that the step lengths are OK but that we are making too many steps in the wrong direction. This could happen, for example, if we use a radially symmetric proposal distribution while the posterior that we are aiming for has strong correlations between parameters, i.e. is ridge-shaped. Besides using a different MCMC algorithm (e.g. an adaptive one, see the first section of this chapter), you can begin by performing some trial-and-error calibration optimisation to find an efficient proposal distribution. Gelman et al. (2013) give some suggestions for optimising the proposal in algorithms like Metropolis that use a multivariate normal distribution to generate the random steps. They advise to make the proposal distribution proportional to the posterior distribution that we are going to generate. However, we do not know this distribution beforehand—that is why we do the MCMC in the first place—so we have to work with the prior distribution instead. Some degree of trial-and-error remains necessary. The most general recommendation that can be made is to try out various proposal distributions in short chains and then select the best for the definitive Bayesian calibration.

2. What do I do if the MCMC crashes? Obviously, this is not supposed to happen! Fortunately it is also very unlikely given the simplicity and robustness of most MCMC algorithms. However, there is one part in the algorithms that occasionally gives problems: calculation of the prior probability for a new candidate parameter vector. If the prior distribution is very simple, e.g. multivariate uniform or triangular, calculation of the prior probability density for any new parameter vector poses no problems. However, if we use a distribution like the multivariate normal for the prior, then calculation of parameter vector probability includes a matrix inversion step, where the variance-covariance matrix needs to be inverted. This may run into limitations with the numerical precision of computers. Matrix inversion is a widely-studied issue in applied mathematics and there are robust but complicated algorithms available. However, we can generally avoid all problems if we don't have too many orders of magnitude difference between the smallest and largest parameters. A large variance-covariance matrix where some entries are in the order of 10^{-6} and others 10^8 is difficult to invert. We can deal with this by rescaling the model's parameters to be all close to unity. Of course, that requires undoing the rescaling each time we need to run our model.

3. What do I do if the MCMC does not converge? The first thing to do is make the chain longer. If that takes too much computing time, consider changing the proposal algorithm, e.g. by making the proposal distribution narrower (i.e. shortening the average proposal step). You can of course also try a different MCMC algorithm altogether.

4. My PBM is too slow for Bayesian calibration because it needs a long spin-up, what do I do? Your model does not *need* a spin-up. Spinning-up, which is pre-running the PBM to equilibrium, is used to initialise the state variables of a dynamic model. This is done to prevent unrealistic model dynamics when the true runs start: the model does not need to find its way back to stable behaviour. However, this generally amounts to replacing initially unrealistic dynamics with unrealistic initial state variable magnitudes, and cannot be recommended. Yeluripati et al. (2009) showed a superior initialisation approach where soil measurements were not discarded but used together with flux-data in a Bayesian calibration of the parameters and initial values of a model for vegetation gas-exchange. This way, spin-up was avoided.

10.5   Results from the Bayesian Calibration

1. What is burn-in and do I need to consider it? Burn-in is the first phase of a MCMC, when the chain is still "searching" for the area of high posterior probability, i.e. before convergence. The parameter vectors sampled during the burn-in phase should be removed from the posterior sample because they are not representative.

2. If my chain length is, say, 10000 and acceptance rate is 25%, does that mean my sample from the posterior only consists of 2500 parameter vectors? No, the sample consists of all 10,000 parameter vectors, but many vectors will appear two or more times in the sample. In other words, if during the MCMC a candidate parameter vector is rejected, the current vector will receive extra weight. So if we want to calculate, for example, the mean posterior parameter vector, we need to sum all 10,000 vectors (including the repeats) and divide by 10,000.

3. What do I do if the sample from the posterior parameter distribution looks wrong, or if the posterior predictions of model output variables look wrong? Not necessarily anything. All the Bayesian calibration does, is combine the information that has been supplied—via the prior, the model and the data-likelihood—and show what that means for what we can say about the parameters. The results may be surprising but nevertheless correct. However, if something really seems amiss, e.g. posterior predictions for some variables seem impossible, there may be an error somewhere. First check the MCMC-code. Then check if the prior was not incorrectly set, e.g. by disallowing perfectly reasonable parameter values. Then check whether the likelihood was correctly defined: perhaps there was less information in the data than you assumed, e.g. if you neglected representation error (where the data were accurate but not representative of what the model actually tries to simulate) or if you overlooked systematic errors in the data. It may also be that you have lots of unused information: if you have good reason to disbelieve certain posterior predictions, then that suggests you have information that could be included in the likelihood function. What's left after all those checks is the most common cause of seemingly poor Bayesian calibration results: the model itself may not be very good and the calibration just makes that very apparent. In that case there are four options left: (1) Accept your results but consider them to be conditional on a model that you think is flawed; (2) Improve your model; (3) Re-interpret your model as not being intended to simulate all the characteristics of the data, e.g. by treating your model outputs as being correct for a specific spatio-temporal scale, which may have lower resolution than that of the data; (4) redo the MCMC with a discrepancy term included (Chap. 13).

4. What do I do with my results, i.e. how do I use my sample from the posterior? Some key characteristics of the posterior to be reported are its mode, mean and variance-covariance matrix. The mode is the vector in the posterior sample with the highest probability, i.e. the highest product of prior and likelihood. It is also called the *Maximum A Posteriori parameter vector* (MAP). The MAP will be equal to the mean if the posterior is symmetric. In most applications of Bayesian calibration, the variance-covariance matrix shows many pair-wise parameter correlations. In many cases the posterior can be well approximated by a multivariate normal distribution, fully defined by the sample mean and variance-covariance matrix. But if the posterior has a more complicated shape advanced *density estimation* techniques may be required. Approximating the posterior sample by a parameterised distribution (i.e. density estimation) is essential in case new data become available with which we want to do a further Bayesian calibration. The posterior from the first Bayesian calibration then becomes the prior for the second Bayesian calibration and we always need a parameterised distribution for the prior to allow us to calculate the prior probability density at any candidate vector in parameter space.

Chapter 11
After the Calibration: Interpretation, Reporting, Visualization

This chapter discusses what needs to be done after your Bayesian calibration: how to interpret your results, what to report and how to report it with emphasis on visualization.

11.1 Interpreting the Posterior Distribution and Model Diagnostics

A major benefit of Bayesian calibration of the model's parameters is that it helps you find structural errors in the model. After calibration, any remaining differences between observations and model outputs will be mainly due to data error and model structural error. It is impossible to fully disentangle these two errors, but some heuristic advice can be given (Van Oijen et al. 2011). We suggest the following four steps, some of which are especially useful for complex process-based models (PBMs):

1. The first step after Bayesian calibration is to inspect the marginal posterior probability distributions, i.e. the probability distributions for each individual parameter in the model. When any of these probability distributions is highly skewed, creeping up to its lower or upper prior bound, then the data contradicted our prior expectations and we need to examine whether there were any errors in prior, data or the model itself. Often it will be the latter—it may be an indication that the process affected by that parameter or related processes are incorrectly implemented. For example, if the posterior litter decomposition rate constant in a soil model is higher than expected, then it may be the case that other decomposition processes were overlooked.

2. The second step is to inspect correlations between parameters. Strong posterior correlations between parameters of any model indicate that some function of the parameters can, a posteriori, be estimated more accurately than the individual

© Springer Nature Switzerland AG 2020
M. van Oijen, *Bayesian Compendium*,
https://doi.org/10.1007/978-3-030-55897-0_11

parameters. A trivial example is where the model is simply the sum of two parameters ($y = \theta_1 + \theta_2$) and where we have one observation of the sum. Then the calibration will induce a strong negative correlation between the parameters, so each individually will be more uncertain than their sum. Other functions of the parameters, such as their difference, may remain highly uncertain (see also Jaynes 2003, p. 267). The presence of strong correlations between parameters may indicate that the model is more complex than it needs to be.

3. The third step is decomposition of the mean squared deviation (MSD) of time series (Kobayashi and Salam 2000). This is of relevance for dynamic models such as PBMs, which for example simulate the dynamics of demographics or biogeochemistry. After the calibration, we can compare observed and simulated time series. The MSD-decomposition proposed by Kobayashi and Salam (2000) splits the MSD into three additive terms that quantify differences between the time series in their mean, variance and phase:

$$MSD = \overline{(z - y)^2} = (\bar{z} - \bar{y})^2 + (\sigma_z - \sigma_y)^2 + 2\sigma_z\sigma_y(1 - \rho),$$

where y refers to observations and z to model output, σ is standard deviations, and ρ is the correlation coefficient. For example, if model and data mean differ (the first term), the model may be missing a process, if the variances differ (second term), the model may be missing a feedback mechanism, if the phases differ (third term), the model may be missing a linked state variable e.g. in a decomposition chain. This MSD-decomposition was applied after Bayesian calibration to the mismatch between observed and simulated gas flux time series for four forest models calibrated for a Norway spruce site in Germany (Van Oijen et al. 2011), to two models for growth and respiration of *E. globulus* in Portugal (Minunno et al. 2013), and to grassland simulations in Germany (Hjelkrem et al. 2017).

4. The fourth step is to check any assumptions of parameter universality. Did we carry out a calibration using data from multiple geographical locations assuming that parameter values should be generic, i.e. the same everywhere? Then we could redo the calibration with genotype- or site-specific parameter estimation. However, this approach pits two extreme ideas against each other: generic versus site-specific parameters. A better approach would be to allow parameters to vary between sites or genotypes but not completely independently—and let the calibration reveal the appropriate degree of parameter independence. Such a more flexible, hierarchical approach is becoming common in environmental and ecological modelling (Cressie et al. 2009) and will be discussed in Chap. 16.

11.2 Reporting

When writing a publication involving Bayesian calibration, we should report on every step. To remind you of these, here is a short checklist that you could use:

1. The data

   - What data did you use? What were the key observational uncertainties?

2. The model

   - What model did you use? If too large for detailed description, where can we find more information about it? How many state variables and parameters does it have?

3. The prior

   - Did you use all parameters or a subset? What joint probability distribution did you assign? What information did you use to quantify it: literature, personal judgment, expert elicitation? Did you use MaxEnt?

4. The likelihood

   - What likelihood function did you assign? How did you account for stochastic, systematic and representativeness errors? Did you treat the model as given or allow for a discrepancy term?

5. The MCMC

   - Which algorithm did you use? How many chains and iterations? How did you assess convergence?

6. The posterior distribution for the parameters

   - What were posterior modes, means, variances and major correlations? How different was the posterior from the prior?

7. The posterior predictive distribution

   - How well did the model, when using the posterior probability distribution for the parameters, (a) reproduce data used in the calibration, (b) predict data not used in the calibration?

Our theme in this book has been that uncertainties are represented by probability distributions, and the checklist above asks to report on them. This raises the issue: how to communicate probability distributions to others in, for example, a scientific publication or a report for management? This is a complicated issue, exacerbated by the fact that the posterior predictive uncertainty for different output variables is typically represented by a high-dimensional joint probability distribution with many non-zero correlations.

Formally, only the full joint probability distribution captures all information about predictive uncertainty. However, summarization is inevitably required, and the way to report uncertainty should depend on both the shape of the distribution and the needs of the persons who are expected to read your report (Spiegelhalter 2011). If the distribution is symmetrical and bell-shaped, then providing the mean value and the standard deviation may be sufficient. For more complicated distributions, reporting quantiles may be preferable and strong correlations should be flagged (see e.g. Van Oijen et al. 2005).

11.3   Visualising Uncertainty

We note that the field of uncertainty communication is still in flux, and several researchers recommend the use of different graphical representations of uncertainty over verbal or numerical reporting (Gabry et al. 2019; Milne et al. 2015; Spiegelhalter 2011). Examples of reporting posterior uncertainties following Bayesian calibration of vegetation models are provided by, amongst others, Van Oijen et al. (2011, 2005; 2013) and a map showing how predictive uncertainty of forest growth varies spatially across the UK is provided by Van Oijen and Thomson (2010).

In its simplest form, the Bayesian method formalizes how information flows from data to model parameters and ultimately model output. In later chapters, we shall see that the method can be expanded to a much more comprehensive framework involving multiple models, model emulators, uncertain drivers, uncertain parameters and hyperparameters combined in a Bayesian hierarchical model. Whilst powerful, the complexity of the framework may obscure a reader's view on the information flows between the different components. In this book, we generally opt for verbal descriptions, mathematical equations and computer code to explain the methodology, but a visual alternative exists in the form of graphical models (GM; e.g. Smith et al. 2012). GMs typically depict all uncertain quantities in separate circles, linked by arrows representing conditional probabilities. For example, an arrow from a circle labelled θ to one labelled y would represent the likelihood $p[y|\theta]$ (see Chap. 15). The use of graphical models is a means of visualizing the Bayesian approach to uncertainties, but equations remain essential to communicate the exact details of the approach. In any case, many different types of graph can be used. The online R Graph Gallery (https://www.r-graph-gallery.com/) provides hundreds of examples.

As a final note, this chapter is titled "After the calibration ..." but visualization can of course help you define likelihoods and find inconsistencies in your prior as well. Gabry et al. (2019) advocate standardizing visualization across the whole Bayesian workflow.

Chapter 12
Model Ensembles: BMC and BMA

12.1 Model Ensembles, Integrated Likelihoods and Bayes Factors

In this chapter, we discuss how multiple 'competing' models can be used simultaneously. There are advantages to having multiple different models, as was already recognized by Chamberlin in the 19th century (Chamberlin 1890). His still highly readable and important essay on "The method of multiple working hypotheses" warned scientists against "parental affection for a favourite theory". He worried that inevitable bias in favour of one's own ideas would lead to "unconscious pressing of the theory to make it fit the facts, and a pressing of the facts to make them fit the theory." And that is —or should be—a legitimate concern for modellers nowadays as well. Moreover, different models can have complementary strengths and we often have no clear idea which of the available models is the best for a given research question. So how can Bayesian thinking help with these issues? Well, as you will expect, the proper Bayesian approach is to quantify our uncertainty about model structure.

Uncertainty with respect to model structure itself is more difficult to represent formally than parameter uncertainty. For parameters, as we have seen, we can simply define a probability distribution $p[\theta]$, but how would we specify a probability distribution $p[f]$? A practical solution to this issue is to collect multiple models and assume that the set forms a representative sample from 'model space'. Such *ensemble modelling* is becoming increasingly common in the environmental sciences (Chandler 2013), and Van Oijen et al. (2013) and Rollinson et al. (2017) provide examples for vegetation models. In ensemble modelling, we define $p[f]$ as a discrete probability distribution over the models in our set. There are then two ways to proceed, based on how we treat model structural error (or 'discrepancy'). The most principled approach is to explicitly include the discrepancy in the analysis and try to quantify it. That will be discussed in the next chapter (Chap. 13). Alternatively, we may make the additional assumption that one of the models in our set is exactly correct, i.e. has no discrepancy at all. That simplifies the treatment considerably because we can then

M. van Oijen, *Bayesian Compendium*,
https://doi.org/10.1007/978-3-030-55897-0_12

find data and use Bayes' Theorem to reduce the uncertainty about which model is correct:

$$p[f|y] \propto p[f]p[y|f], \tag{12.1}$$

where the model likelihood, $p[y|f]$, requires integrating out parameter uncertainty:

INTEGRATED LIKELIHOOD:

$$p[y|f] = \int p[y|\theta, f]p[\theta]d\theta. \tag{12.2}$$

We refer to $p[y|f]$ as the *integrated likelihood* for model f. It is, as the equation shows, the expectation of the likelihood value under the prior probability distribution for the parameters. Let's consider how we would apply these equations to model comparison. If we have two models f_1 and f_2 and no prior preference for either (i.e. $p[f_1] = p[f_2] = 1/2$), then we can use data y to calculate the posterior odds of f_1 as the ratio of the integrated likelihoods: $\frac{p[f_1|y]}{p[f_2|y]} = \frac{p[y|f_1]}{p[y|f_2]} = \frac{\int p[y|\theta_1, f_1]p[\theta_1]d\theta_1}{\int p[y|\theta_2, f_2]p[\theta_2]d\theta_2}$. That ratio of integrated likelihoods is called the *Bayes Factor* and gives us the formally correct way to use data for updating model odds (Kass and Raftery 1995).

12.2 Bayesian Model Comparison (BMC)

We thus find that the key quantity in *Bayesian Model Comparison* (BMC) is each model's integrated likelihood. The fact that it is calculated by integrating over a model's whole parameter space means that it has an in-built 'Occam's razor' that favours model simplicity (MacKay 1992). Parameter-rich models may achieve high likelihood in some small part of their parameter space, but the part of parameter space where they perform poorly will be much larger than for lower-dimensional models. Parameter richness will thus only confer high integrated likelihood if the model's parameter distribution restricts its parameter values to the parts of parameter space where the model fits the data well. In practice this means that complex models will only be preferred if sufficient data have been available to calibrate their parameters well. So when comparing two or more models, there is no need to penalize parameter richness in some ad-hoc way: Bayes' Theorem will take care of that automatically through the integrated likelihood. Jaynes (Jaynes 2003: Fig. 4.1) gives a nice example of BMC in which the relative probabilities of three hypotheses change with gradually accumulating evidence. An advance warning though: reliably estimating the integrated likelihood for a model is difficult if parameter space is high-dimensional (Kass and Raftery 1995).

When comparing process-based models, there is the difficulty that some models may produce a different suite of output variables than other models do. To keep their integrated likelihoods comparable, we need to restrict ourselves to observations on variables that all the models can predict. Moreover, the data should obviously not already have been used for parameter calibration of any of the models. There is also the issue that some models may have been calibrated much better than others. For example, we may learn that some colleagues have produced a forest model that works very well for the tree species in their country, and we want to compare their model to our own. In that case, it is useful to have new data that can be split into a calibration set and a model comparison set, as was done by Van Oijen et al. (2013) in their BMC of six different process-based forest models. An alternative approach is *Reversible Jump MCMC* (RJMCMC) (Green 1995) that achieves parameter calibration and model comparison by MCMC in one go, using just one data set, by joining the models' parameter spaces together and also adding a 'model indicator parameter' whose discrete posterior probability distribution is the sought after $p[f|y]$. This works well for parameter-sparse models.

One final remark about BMC. The integrated likelihood is the expectation value of the likelihood under the prior distribution for the parameters, so its value depends not only on the data but also on the prior parameter distribution for the parameters. Now imagine that two researchers want to compare their models without knowing that they both have exactly the same model... The only difference would be their 'expertise' concerning plausible parameter values. So they would have different parameter priors, leading to different integrated likelihoods. The researcher with the greatest expertise would probably achieve a higher integrated likelihood than his counterpart—his parameter values will make more sense—and his 'model' would be given the highest posterior probability in the BMC. But how can that be correct, if both models are the same? Well, the answer *is* correct, and confirms that models without a parameter distribution are not completely defined. A model is only as good as its parameter distribution allows it to be. Models cannot be compared on their structure alone, we always evaluate 'model + parameter distribution'.

12.3 Bayesian Model Averaging (BMA)

After we have carried out a BMC, we can account for model uncertainty in our predictions. We make use of the whole model ensemble by weighing each model's contribution to prediction with its posterior probability $p[f|y]$. This is referred to as *Bayesian Model Averaging* (BMA) (Kass and Raftery 1995; Van Oijen et al. 2013). Formally:

BMA after BMC:

$$p[z = \hat{z}|y] = \sum_{i=1}^{n_f} p[f_i|y]\, p[z = \hat{z}|f_i], \tag{12.3}$$

where n_f is the number of models in our ensemble. Note that BMA does not average predictions but probabilities. For argument's sake: if we have two equally plausible models, and their predictions are $z = 10$ and $z = 20$ with complete certainty, then the BMA is not $z = 15$ but $p[z = 10] = 0.5$; $p[z = 20] = 0.5$. So predictive distributions from BMA are usually multimodal. Despite that, their predictive capacity tends to be superior to those of any single model in the ensemble. But more importantly, BMA is the correct way of accounting for the uncertainty embodied in our ensemble, and is therefore to be preferred over just selecting the most plausible model and discarding the others. Let's now look at an example of BMC and BMA in action.

12.4   BMC and BMA of Two Process-Based Models

12.4.1   *EXPOL5 and EXPOL6*

We shall be comparing two versions of the expolinear process-based model (Goudriaan and Monteith 1990). We already presented the first version, EXPOL5, in Chap. 9, where we calibrated the model's parameters by means of MCMC. EXPOL5 is a 5-parameter model that—unusual for process-based models—can be analytically solved. We shall be using the same prior distribution for the parameters of EXPOL5 as in the earlier chapter.

Our second model is EXPOL6 which was also proposed by Goudriaan. It is a different version of the expolinear model but it is also analytically solvable. As you will be expecting, EXPOL6 has 6 parameters rather than 5. The extra parameter is $LAIMAX$, the maximum value of the leaf-area index (LAI). So in this model the LAI does not continue growing proportionally to biomass W, but levels off to an asymptote. Here is the R-function EXPOL6 from which you can infer the mathematical formula yourself. Just like EXPOL5 it generates output for two variates, W and LAI.

```
EXPOL6 <- function( t=0, b=c(10,1,3,0.007,2,1) ) {
   I0  <- b[1] ; K    <- b[2] ; LAIMAX <- b[3]
   LAR <- b[4] ; LUE  <- b[5] ; W0     <- b[6]
   e1  <- exp(I0*K*LAR*LUE*t) ; e2 = exp(K*LAR*W0) ; e3 = exp(-K*LAIMAX)
   W   <- W0*e3 + ( log(e2+1/e1-1)/(K*LAR) + I0*LUE*t ) * (1-e3)
   LAI <- log( (1+e1*(e2-1)) / (1+e1*(e2-1)*e3) ) / K
   return( list( "W"=W, "LAI"=LAI ) ) }
```

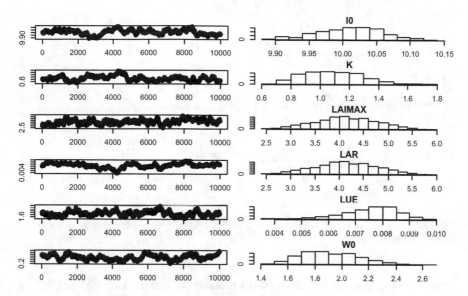

Fig. 12.1 Bayesian calibration of the 6-parameter expolinear model by means of Metropolis sampling

12.4.2 Bayesian Calibration of EXPOL6's Parameters

Here is the prior for `EXPOL6` that we shall be using.

```
bmin6   <- c(  9.9, 0.5, 1, 0.005, 1, 0.5 )
bmax6   <- c( 10.1, 1.5, 5, 0.009, 3, 1.5 )
mb6     <- rowMeans( cbind(bmin6,bmax6) )
range6  <- bmax6 - bmin6 ; var6 <- (range6^2) / 12 ; Sb6 <- diag(var6)
```

In Chap. 9, we calibrated `EXPOL5` against a dataset on biomass W and leaf-area index LAI. To put both models on an equal footing before comparing them, we calibrate `EXPOL6` against the same data.

```
bChain <- MetropolisLogPost( f=EXPOL6, logp=logPostList, dataList=data )
```

We showed the parameter trace plots and parameter histograms for `EXPOL5` in the earlier chapter (Fig. 9.1), so we only show the MCMC-results for `EXPOL6` here (Fig. 12.1). Plots of `EXPOL6` outputs vs. data for the posterior mean parameter vector are in Fig. 12.2).

12.4.3 BMC and BMA of EXPOL5 and EXPOL6

We have defined the two expolinear models and calibrated their parameter probability distributions against the same data. Both models and their posterior parameter distributions shall now be confronted with a new data set. This consists of data on

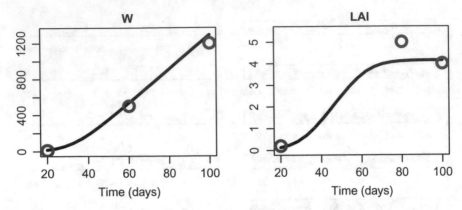

Fig. 12.2 Posterior mean growth curves for model EXPOL6: biomass (W, $g\,m^{-2}$) and leaf-area index (LAI, $m^2\,m^{-2}$)

both of the models' output variables, the biomass W and the leaf-area index LAI. As is typical in practice, we shall not have the same number of data points for both variables. In this case, we have fewer observations for LAI than for W.

```
t          <- c(10,20,40,80) ; y <- c(4.3,17.0,162.0,887.4)
sd         <- c(2,4,20,100)   ; data_W   <- cbind( t, y, sd )
t          <- c(20,80)        ; y <- c(0.13,4.14)
sd         <- c(0.1,0.1)      ; data_LAI <- cbind( t, y, sd )
data_NEW <- list( "W"=data_W, "LAI"=data_LAI )
```

We first calculate the integrated likelihood EXPOL5 using the new data set. We do this by taking a large sample of parameter vectors from the joint probability distribution for the model's five parameters, calculating the data-likelihood for each parameter vector, and taking the average. The following code essentially does that, although it works with log-transformed probabilities to make the calculation more numerically stable.

```
ns             <- 1e3
sample_b5      <- rmvnorm( n=ns, mean=mb5_y, sigma=Sb5_y )
sample_log_L5 <- sapply( 1:ns, function(i) {
  logLikList( sample_b5[i,], dataList=data_NEW, f.=EXPOL5 ) } )
log_IL5        <- mean(sample_log_L5) +
                  log( mean( exp(sample_log_L5-mean(sample_log_L5)) ) )
```

Note that we resampled the posterior parameter distribution from the multivariate normal with the posterior mean and covariance matrix that we found in Chap. 9. This assumes that the posterior is well represented by that normal distribution. We could instead have directly subsampled from the MCMC chain that we generated when calibrating EXPOL5 on the first data set, and that is generally the more reliable method. We now carry out the same sampling for EXPOL6 to calculate the integrated likelihood of that model.

After this is done, and the log-transformation is reversed, we find that the two integrated likelihoods are:

- $p[y|EXPOL5] = 1.428 \times 10^{-6}$,
- $p[y|EXPOL6] = 1.313 \times 10^{-6}$.

So the posterior probabities are:

- $p[EXPOL5|y] = 0.5209$,
- $p[EXPOL6|y] = 0.4791$.

Finally, these results allow us to use BMA for prediction. Say we want to predict $W(120)$, the biomass at time 120. The proper procedure would then be to (1) propagate parameter uncertainty for both models to gives us two predictive probability distributions, (2) mix the two distributions according to the posterior probabilities that we just found.

Chapter 13
Discrepancy

We have seen that if we have a probability distribution for our model's parameters, then we can sample from that distribution to see how parameter uncertainty translates into predictive uncertainty. And if we get new data, then we can use Bayes' Theorem to update the parameter distribution and thereby reduce our predictive uncertainty. So far, so good. But a more difficult problem is that of uncertainty about model structure. We know that all models are wrong, but not how wrong they are. In the preceding chapter on Bayesian Model Comparison (Chap. 12), we addressed the issue of model structural uncertainty by collecting an ensemble of models, rich enough that we could believe it contained the 'correct' model, followed by using data to help us quantify the relative plausibilities of the different models. But what if we are not convinced at all that our ensemble contains a really good model, let alone a correct model? And what if we have only one model that we can work with? In such cases, we need to tackle the issue of model deficiency, or *discrepancy*, head-on.

Let's take a step back and discuss the inevitable limitations of scientific modelling. We can define *theory* as the collection of generally accepted causal relationships in a scientific discipline. In well-developed fields, such as engineering and architecture, effective models can sometimes be derived from theory alone. In other fields, such as medicine, theory is often lacking. The environmental and ecological sciences are somewhere in between. We have theory about many components of the earth system, but not enough to derive unique models from. Therefore a wide range of modelling approaches is still in use, ranging from simple statistical models to complex process-based ones, as we see in this book. But whatever model we use, it is only an approximation to nature, so it is strictly speaking an incorrect representation of nature. To quantify this structural error, we use the term *discrepancy* (Chandler 2013; Rougier 2007). It is defined as the model error that remains when all parameters have been set to the best possible value. So the discrepancy is the difference between model behaviour for optimum parameter values, and the true behaviour of

© Springer Nature Switzerland AG 2020
M. van Oijen, *Bayesian Compendium*,
https://doi.org/10.1007/978-3-030-55897-0_13

the natural system. Unfortunately, we never know exactly how large the discrepancy is. If we did, we would modify our model. Our uncertainty about the discrepancy complicates our attempts at Bayesian calibration (Kennedy and O'Hagan 2001).

13.1   Treatment of Discrepancy in Single-Model Calibration

As explained, Bayesian calibration combines prior information with the likelihood, and it is in the latter term that model outputs are compared with data. There are three different ways in which the discrepancy can be treated in the likelihood: (1) ignore it, (2) lump it with data error, (3) account for data error and discrepancy separately. We now discuss the merits of the three methods.

First, we may choose not to represent the discrepancy at all, in which case our calibration is formally conditional on the model being correct. For ease of exposition, let us assume that error uncertainties are represented by zero-mean Gaussians. Then our likelihood function would read:

$$p[y|\theta] = N[y - f(x, \theta); \mu = 0, \sigma^2 = \sigma_y^2], \qquad (13.1)$$

where we have the usual choice of pre-specifying the value of σ_y^2 or adding it to the parameters to be calibrated, in which case we must specify a prior distribution for it. Ignoring the discrepancy in this way is quite common; it is akin to linear regression without considering the possibility of a nonlinear function. It is a simple method that underestimates predictive uncertainty of the model because it can only propagate input uncertainty, not structural uncertainty. This method has been used repeatedly in vegetation modelling (e.g. Van Oijen et al. 2005), but we may want to move away from it.

The second approach also does not recognize the discrepancy explicitly but estimates the sum of data and model structural error, i.e. the overall mismatch between data and model outputs. Again using Gaussians, the likelihood would then read:

$$p[y|\theta] = N[y - f(x, \theta); \mu = 0, \sigma^2 = \sigma_{mismatch}^2], \qquad (13.2)$$

where we make no attempt to decompose $\sigma_{mismatch}^2$ into contributions from data error vs. discrepancy. In this method, the $\sigma_{mismatch}^2$-term cannot be pre-specified—because we cannot foresee how much model outputs will deviate from data—and must be added as a parameter to be calibrated. This method was used by, amongst others, Kavetski et al. (2006). Despite being simple, this method is also not to be advised, as it obscures the difference between errors in data and model (Rougier 2007). It also leads to overestimation of predictive uncertainty: we do not want to predict future measurements ($f(x, \theta) + \epsilon_{mismatch} = f(x, \theta) + \epsilon_y + \epsilon_{model}$) but future true values ($f(x, \theta) + \epsilon_{model}$).

This leaves the third method, rarely used but to be recommended: explicitly distinguishing errors in data from those in model structure:

$$p[y|\theta] = N[y - f(x, \theta); \mu = 0, \sigma^2 = \sigma_y^2 + \sigma_{model}^2], \qquad (13.3)$$

where we have to specify a prior for σ_{model}^2. In this approach, we can account for all sources of model predictive uncertainty ($p[x]$, $p[\theta|y]$, $p[\epsilon_{model}]$). Although this method is in principle superior to any other, it may be difficult to specify the prior for the discrepancy because simple Gaussians (or Gaussian Processes, see Chap. 14) are unlikely to be adequate. However, the method has been sketched out in ground-breaking papers (Kennedy and O'Hagan 2001; Rougier 2007) and should be developed further. Brynjarsdóttir and O'Hagan (2014) provide a very clear discussion of the need for discrepancy modelling and its difficulties.

13.2   Treatment of Discrepancy in Model Ensembles

So far, we have discussed two ways of taking into account that every single model is wrong: ensemble modelling in Chap. 12 and discrepancy quantification in this chapter. Let's now attempt to combine both methods. We start with the conceptual diagram shown in Fig. 13.1.

The figure shows the behaviour of a system in reality as well as the behaviour of models. The figure is a map, with every map location representing a different trajectory of the system state, i.e. a time series for the multivariate system state of interest. So despite the figure being two-dimensional, every point represents an ultra-high dimensional quantity. On the left we see reality but the models are on

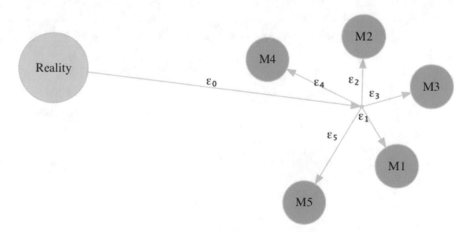

Fig. 13.1 Conceptual diagram for model discrepancy, showing model-specific biases and an overall bias shared by the whole cluster. Simplified after Chandler (2013)

the right, implying that all of them are wrong. Each model's distance from reality measures its discrepancy. The models are clustered, so to some degree they share a common bias. That means we can decompose each model's discrepancy as the sum of a shared bias and a model-specific deviation from the centre of the cluster. The figure is a simplification of a scheme proposed by Chandler (2013) for climate model ensembles, based on the realisation that there is always some convergence in model development, leading to structural errors that are shared by all models in the ensemble. Chandler and colleagues suggest that we try to quantify the two components of discrepancy (ensemble bias and model-specific deviation) separately (Spence et al. 2018). A similar decomposition of discrepancy was proposed by Goldstein et al. (2010) who used the terms *internal* and *external structural uncertainty*. Goldstein provided suggestions for the kinds of model structural and parametric sensitivity analysis that a modeller could perform to quantify internal uncertainty, and the more informal expert elicitation needed to estimate external uncertainty (Goldstein and Rougier 2009). Overall, this approach seems highly promising. It accounts, in principle, for model structural uncertainty in the most comprehensive and realistic way. However, it remains hard to make all steps fully operational. If we want to estimate the scheme components in a Bayesian way, then we need to assign prior probability distributions to all biases (for ensemble mean and for individual models), but we generally have no heuristics telling us how to do so. At the very least, Fig. 13.1 can keep us on our toes by reminding us that convergence of models does not imply convergence to the truth!

Chapter 14
Gaussian Processes and Model Emulation

14.1 Model Emulation

Sampling-based estimation of the posterior distribution is computationally demanding. We have already mentioned the continuing search for efficient MCMC algorithms. MCMC is especially slow when the model of interest is a process-based
model (PBM) with a long run-time. In such cases it may be good to replace the PBM
with a faster surrogate model. The surrogate model will take the same inputs as the
original model, but calculate the output more quickly. However, its output cannot be
exactly the same as that of the original model, so it just provides an approximation.
If the surrogate model is a statistical model that produces not just the approximative
prediction of what the original model would have produced, but a whole probability
distribution, then it is called a *statistical emulator*, or just *emulator* for short.

So an emulator is a replacement for the original model f. It is also a function of
x and θ, but its output is not $f(x, \theta)$ itself but a probability distribution for it. So
the emulator is a tool for predicting—probabilistically—the output from our model
f without actually running the model. The idea is of course that we can build the
emulator in such a way that it will be much faster to evaluate than the original model.
The emulator $p_g[..]$ is usually derived from a small training set of N modelling results
$\{x_i, \theta_i, f(x_i, \theta_i)\}_{i=1:N}$:

$$p_g[\, g \mid x, \theta \,] \;=\; p[\, f(x, \theta) = g \mid \{x_i, \theta_i, f(x_i, \theta_i)\} \,]. \qquad (14.1)$$

Because emulators are designed to compute faster than the original model, they are
especially useful when we need a large sample of model runs. Bayesian calibration
(using MCMC) is one example where we have to evaluate a model many times to
create a representative sample from the posterior distribution. However, emulator
speed comes at the cost of only producing a probabilistic answer. This makes the
likelihood function more complicated. Say our uncertainty about measurement error
is captured by the parameter σ_y. Then the likelihood function for the original model

© Springer Nature Switzerland AG 2020
M. van Oijen, *Bayesian Compendium*,
https://doi.org/10.1007/978-3-030-55897-0_14

would be $L[\theta] = p[\epsilon_y = y - f(x, \theta)|\sigma_y]$, but when using the emulator we would have to integrate out our uncertainty for g:

$$L'[\theta] = \int_{g=-\infty}^{\infty} p[\epsilon_y = y - g|\sigma_y] \, p_g[g|x, \theta] \, dg. \tag{14.2}$$

In Bayesian calibration, the posterior would then be proportional to $p[\theta]L'[\theta]$. To keep calculation of the likelihood-integral as well as $p[\theta]L'[\theta]$ manageable, emulators are generally selected for their mathematical properties. In particular, Gaussian process emulators are often used (Kennedy and O'Hagan 2001), which approximate the original model using the multivariate normal distribution and use convenient functions for calculation of covariances.

14.2   Gaussian Processes (GP)

A *Gaussian Process (GP)* is a multivariate Gaussian distribution defined for infinitely many variates. Let's track back a bit here to explain what that means. The univariate Gaussian captures our uncertainty about a single quantity, and it is fully defined by just two parameters: its mean and variance. The bivariate Gaussian represents uncertainty about two quantities, so its definition requires us to quantify five parameters: two mean values, two variances and one covariance. A trivariate Gaussian has nine parameters: three mean values, three variances, three covariances. We see that the number of parameters for n-variate Gaussians increases as $n(n + 3)/2$. So how can a GP, defined as a Gaussian where n is infinite, be of any practical use? Well, the solution is to define two functions: a *mean function* $m(s)$ and a *covariance function* $C(s, s')$ where the s and s' range over the infinite set of quantities. Note that variances are also covered by the covariance (or 'kernel') function because a variance is just the covariance of a quantity with itself ($C(s, s)$).

When do we use a GP? The typical case is when we are uncertain about a function $f(s)$, i.e. we do not know the values of $f(s)$ for most or all of the values of its arguments s. We can have that uncertainty because the function is nature itself, e.g. when we are interested how, in a region with spatial coordinates $s = (latitude, longitude)$, some soil property $f(s)$ varies. Our knowledge of soil properties and our understanding of soil properties will often be incomplete in which case we may decide to represent our uncertainty about soil property variation as a GP. Using a GP for such problems is called *kriging*, but it can also be considered a form of *machine learning* (see Chap. 20). GP can likewise be used if f is a model rather than nature itself. Say we are uncertain about the outcomes of a computer model f across parameter space $\{\theta \in \Theta\}$, and the model is too slow to run for many values of θ. Expressing our uncertainty about model outputs $f(s)$ using a GP is called model *emulation*, which is the subject of this chapter. Finally, the $f(s)$ can involve both nature and modelling. For example, we may use a GP to represent our uncertainty about how much the model outputs differ from reality for different model inputs (covariates x and/or parameters θ). In such a case, we are using a GP to represent the *discrepancy*. Other possible uses of GP are for representing variation in measurement error or model-data mismatch as a function of covariates: the possible uses of GP are endless.

Let's summarize. We denote a GP for a function of interest f as $GP[m(s), C(s, s')]$, often shortened as $GP[s]$, where $m(s) = E[f(s)]$ and $C(s, s') = Cov[f(s), f(s')]$. The GP is a multivariate normal distribution that is fully defined by is mean function $m(s)$ and its covariance function $C(s, s')$, defined for all s and s' in the domain S. The following two equations are then equivalent ways of expressing that we use a GP-emulator to represent our uncertainty about model output $f(s)$:

EMULATION USING A GAUSSIAN PROCESS:
$$f(s) \sim GP[m(s), C(s, s')] \tag{14.3}$$
$$f(s) = m(s) + T(s), \quad \text{where } T(s) \sim GP[0, C(s, s')].$$

For more details about GP and their connections with various statistical methods, see Kennedy and O'Hagan (2001) and Conti and O'Hagan (2010).

It is common to build emulators for *scalar* output from models, e.g. the final yield or average growth rate. So if the model that we want to emulate generates predictions for n different variables, then we could decide to build n different emulators, one for each of the outputs. And if each output variable comes as a predicted time series, then each individual emulator could represent, for example, the time-averaged output for a single variable (Andrianakis et al. 2015). For time series or spatially gridded output, an index of time or space could be added to the inputs of the emulator (Kennedy and O'Hagan 2001). For another (Bayesian hierarchical) approach to combining scalar emulators to represent spatio-temporally distributed model output, see Leeds et al. (2013).

The more elegant—but difficult to implement—idea of directly emulating *multivariate* output from models is a matter of ongoing research (Conti and O'Hagan 2010; Drignei 2017; Higdon et al. 2008; O'Hagan 2006; Rougier 2008). And for other approaches than GP in emulation, see e.g. Hauser et al. (2012).

A note of caution about the literature. Earlier in this book, we objected to the term *random variable* because nothing is really random in nature. There is only our own uncertainty, and what we represent by means of probability distributions is just that—our uncertainty—and not some intrinsic randomness. The terminological inconsistency is most stark when modellers define their model parameters to be *constants* yet when they try to estimate their values consider them to be random *variables*! Likewise, we should object to the use of the term *random function* to denote a function whose behaviour we are uncertain about. This is just a heads-up for when you start studying the literature on emulation, where an emulator is often described as being a random function.

14.3   An Example of Emulating a One-Input, One-Output Model

We shall now be building a GP-emulator for a very simple model, which has only one input variate and one output variate. So our model is a function f that takes scalar input x_i and provides one matching scalar output $y_i = f(x_i)$. Let's further assume that we have run the model for three values of x_i, and we have collected the inputs and outputs in vectors $x = \{x_i\}$ and $y = \{y_i\}$ for $i = 1..3$. Let's now use these training data to build a GP emulator of the model, and use the emulator to estimate the function for values of x that were not evaluated.

```
x <- c(10,20,30) ; y <- c(6.09,8.81,10.66)
```

We start with the simple case where we have values for all GP parameters except the regression parameters β. In other words, we have the covariance function of our GP, but not its mean function. For the covariance function in this example, we choose an overall variance $V_y = 3$, and we assume that the covariance between points decreases as a negative exponential function of distance with a correlation length $\phi = 10$. So we find the covariance matrix for the training data with the following code:

```
Vy <- 3 ; phi <- 1e1      # GP variance and correlation length
dx <- as.matrix( dist(x) ) # Distance matrix
Sy <- exp( -dx/phi ) * Vy  # Covariance matrix
```

That only leaves the prior for the regression parameters to be specified. We have two regression parameters (an intercept and a slope), so the design matrix X for the regression will consist of two columns: one filled with all 1's, the other with x. For the parameter-pair we specify a prior mean vector μ_β and a prior covariance matrix Σ_β, which we denote in R-code as mb and Sb.

```
mb <- c( 0, 0 ) ; Sb <- diag( 1e4, 2 )
```

Just to remind you of how the design matrix X works: the matrix-vector product $X \mu_\beta$ is a vector consisting of the prior regression values for all model input values x. At this point, we can choose two ways to continue: we can implement the GP for this model ourselves using known analytical solutions, or we can use an R-package to build the GP for us. We show both ways, but will check that they give the same results. The R-package (we choose geoR) will turn out to be useful both for producing plots and for trying out other covariance functions than our main choice, the negative exponential. But let's begin with showing the analytical solutions explicitly.

14.3.1   Analytical Formulas for GP-calibration and Prediction

When, as in this example, the covariance matrix Σ_y has been assigned (Sy in the R-code), the posterior distribution for the regression parameters can be found analytically (Lindley and Smith 1972) as we already showed in Chap. 8:

GP with only regression parameters unknown:

$$p[\beta|y] = N[\mu_{\beta|y}, \Sigma_{\beta|y}], \text{ where}$$

$$\Sigma_{\beta|y} = (\Sigma_\beta^{-1} + \mathbf{X}^\top \Sigma_y^{-1} \mathbf{X})^{-1},$$

$$\mu_{\beta|y} = \Sigma_{\beta|y}(\Sigma_\beta^{-1}\mu_\beta + \mathbf{X}^\top \Sigma_y^{-1} y). \qquad (14.4)$$

These equations can be programmed as follows:

```
GP.est <- function( x, y, Sy, mb, Sb, X=cbind(1,x) ) {
  Sb_y <- solve( solve(Sb) + t(X) %*% solve(Sy) %*% X )
  mb_y <- Sb_y %*% ( solve(Sb) %*% mb + t(X) %*% solve(Sy) %*% y )
  return( list( "mb_y"=mb_y, "Sb_y"=Sb_y ) ) }
```

And with our choices for data and prior this gives the following results for the posterior mean and covariance matrix of the regression parameters of this GP:

$$\mu_{\beta|y} = \begin{bmatrix} 3.9068 \\ 0.2286 \end{bmatrix}; \quad \Sigma_{\beta|y} = \begin{bmatrix} 6.742 & -0.259 \\ -0.259 & 0.013 \end{bmatrix}. \qquad (14.5)$$

At this point, we have fully defined and trained our GP. Our mean function is $X\beta$ with parameter uncertainty $N[\mu_{\beta|y}, \Sigma_{\beta|y}]$ and our covariance function is the negative exponential $C[s, s'] = V_y \exp(-d(s, s')/\phi)$ with $V_y = 3$ and $\phi = 10$. So we can now study our model function f, which is unknown apart from the training data, by means of this emulator instead of f itself.

But how exactly do we use the emulator to predict the output from f for a point x_0 where we do not have the function evaluation $f(x_0)$? Well, because the GP is a well-behaved multivariate normal distribution, the predictive distribution $N[\mu_{y_0|y}, \Sigma_{y_0|y}]$ can be found analytically too. Essentially it boils down to: (1) finding the distances from x_0 to all training locations $\{x_i\}$, (2) using the distances to calculate the covariances $\{C(x_0, x_i)\}$, and (3) using the covariances to determine how much $\mu_{y_0|y}$ and $\Sigma_{y_0|y}$ differ from, respectively, $X_0\beta$ and V_y. The R-code for the analytical solution is as follows:

```
GP.pred <- function(x0,x,y,Sy,phi,mb_y,Sb_y,X0=c(1,x0),X=cbind(1,x)) {
  dx0  <- if( is.vector(x) ) { abs(x-x0)
         } else { sapply(1:length(y),function(i){dist(rbind(x0,x[i,]))}) }
  C0   <- Sy[1] * exp( -dx0/phi )
  m0_y <- X0 %*% mb_y - t(C0) %*% solve(Sy) %*% (X %*% mb_y - y)
     a <- X0 - t(C0) %*% solve(Sy) %*% X
  S0_y <- Sy[1] - t(C0) %*% solve(Sy) %*% C0 + a %*% Sb_y %*% t(a)
  return( list( "m0_y"=m0_y, "S0_y"=S0_y ) ) }
```

[The ugly if-else construction that we use in this code to calculate the vector of distances to the target point x0 allows the function also to be used when there is more than one coordinate, i.e. when x is not a vector but a higher-dimensional array. This will be useful later.]

Before we start using these predictive equations, note that for both the posterior mean function and the posterior covariance function, we need to invert the prior covariance matrix Σ_y (Sy in the R-code). So when we choose a covariance function for our GP, we do not have complete freedom: we must choose a function that leads to a proper invertible symmetric covariance matrix. In mathematical parlance, we must ensure that the matrix is 'positive semidefinite' (Rasmussen and Williams 2006). But that still leaves much choice, such as exponential, Gaussian (squared exponential), spherical and other covariance functions. We shall explore a few of these functions later in this chapter.

Let's now calculate results of our emulator for the as yet unevaluated points $x_0^a = 10.1$ and $x_0^b = 15$. The first is very close to a point in our data set ($x[1] = 10$), the second is not close. Our R-function GP.pred gives the following results for predictive mean and variance at these two points:

```
> x0=10.1: mean = 6.117023 ; variance = 0.05926102
> x0=15.0: mean = 7.437081 ; variance = 1.410466
```

As expected, the predictive mean for the first point is very similar to the nearby observation value ($y[1] = 6.09$) and it has low uncertainty, whereas the second point has a mean predictive value that is roughly the average of its nearest neighbours, but those are not nearby so the predictive variance is very high.

Notice that in the example that we have just given, the x played a double role, as coordinates and as covariate-values. We used the x as coordinates to calculate distances between points and from those distances we calculated the covariances. And we used the x as covariate-values in our regression. We shall often see GP with this double role of x, especially in emulation. But it is quite possible to have a GP where the x are coordinates and nothing more, e.g. when we just want to estimate a global mean without any regression on covariates. In that case the covariate-matrix X would just consist of a single column with the value 1 in each position, and there would be no columns with coordinate values. And on the other hand, it is also possible to have a GP where for each location we have more covariates than coordinates, e.g. in some forms of geostatistical kriging.

14.3.2   Using R-Package *geoR* for GP-calibration and Prediction

The R-package 'geoR' requires specifying two spatial coordinates, typically longitude and latitude, but when we specify the same latitude for all points we mimic a one-input model, which is what we need here for our simple univariate model. We now use the package to find the posterior for the intercept and slope, and then calculate the marginal predictive distributions for the same two points as above (one close to a measured point, and one further away). Here is the R-code:

```
s.11      <- as.geodata( cbind( x, rep(0,3), y) )
xpred     <- c(10.1,15) ; xpred.11 <- expand.grid( xpred, 0 )
model.11  <- model.control(cov.m  ="exponential",
                           trend.d=~coords[,1], trend.l=~ xpred.11[,1])
prior.11  <- prior.control(beta.prior    ="normal", beta     =mb,
                           beta.var.std=Sb/Vy,
                           sigmasq.prior ="fixed" , sigmasq  =Vy,
```

```
                              phi.prior       ="fixed" , phi      =phi,
                              tausq.rel.prior="fixed" , tausq.rel=0)
out.11    <- output.control(messages=F)
geoR.11   <- krige.bayes(s.11,loc=xpred.11,mod=model.11,pr=prior.11,out=out.11)
```

The posterior parameter distribution found by geoR is the following:

```
> Mean: 3.906838 0.228601
> Variance:
>                  beta0         beta1
> beta0  6.7424941
> beta1 -0.2592242 0.01296323
```

The parameter estimates from geoR are, as we would hope, the same as we found with the direct implementation of the theoretical equations. And now let's examine geoR's prediction for $x_0^a = 10.1$ and $x_0^b = 15$:

```
> Means: 6.117023 7.437081 , Variances: 0.05926102 1.410466
```

The prediction results are also the same as we found analytically. So we can rely on geoR whenever that is more convenient than implementing the theory ourselves. Let's now conclude this section by using geoR to calculate the predictive distribution for many more points. We can then plot these to show the shape of the emulated function across its domain, with uncertainty bounds. We will first do so using the exponential covariance function that we have used all the time up to now, but we shall also show the consequences for the GP interpolation of using ten different covariance functions. The results are shown in Fig. 14.1.

We see that the choice of covariance function generally matters little for the posterior mean-function (the unbroken lines in the figure) but has major impact on

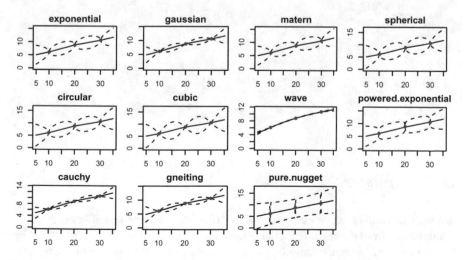

Fig. 14.1 GP emulation of a 1D model y=f(x) by means of geoR: results for 11 different covariance functions

our predictive uncertainty away from training points. Note also that the last panel differs from the others. It shows the results from geoR's 'pure.nugget' covariance function, which actually represents no spatial covariance at all, just local noise. If you choose any of the other covariance functions but let the spatial correlation length ϕ approach zero, your results will be the same as for "pure.nugget".

14.4   An Example of Emulating a Process-Based Model (EXPOL6)

The remainder of this chapter will be concerned with GP-emulation of model EXPOL6. This is a 6-parameter version of the expolinear model (Goudriaan and Monteith 1990) that we introduced and calibrated in Chaps. 9 and 12. The model produces output of biomass W and leaf-area index LAI as a function of time t and 6 model parameters. We want to make an emulator for the LAI-predictions. For the emulation, we shall only use our own equations as geoR cannot go beyond two dimensions as far as the parameter covariance matrix is concerned.

We shall be working with the same prior distribution for the model that we specified in Chap. 12. To make the GP-calculations computationally robust, we use a 'scaled' version of the model, which we call EXPOL6s, where the magnitude of every parameter is around 1. Within the R-definition of this modified model, the dimensionless scaled parameter values are transformed back to their original units. Note from this R-code that also the covariate t (which represents time and is the first input to the model) is scaled, in its case by division by 100.

```
EXPOL6s <- function( t=0, b=c(1,1,1,1,1,1) ) {
   t        <- 100    * t
   I0       <- 10     * b[1]  # MJ PAR m-2 ground d-1
   K        <- 1      * b[2]  # m2 ground m-2 leaf area
   LAIMAX   <- 3      * b[3]  # m2 leaf area m-2 ground
   LAR      <- 0.007  * b[4]  # m2 leaf area g-1 DM
   LUE      <- 2      * b[5]  # g DM MJ-1 PAR
   W0       <- 1      * b[6]  # g DM m-2 ground
   e1       <- exp(I0*K*LAR*LUE*t) ; e2 = exp(K*LAR*W0) ; e3 = exp(-K*LAIMAX)
   W        <- W0*e3 + ( log(e2+1/e1-1)/(K*LAR) + I0*LUE*t ) * (1-e3)
   LAI      <- log( (1+e1*(e2-1)) / (1+e1*(e2-1)*e3) ) / K
   return( list( "W"=W, "LAI"=LAI ) ) }
```

14.4.1   Training Set

We need a training set of completed EXPOL6s calculations to calibrate our GP-emulator on. So we sample from the prior parameter distribution for EXPOL6s by means of *Latin Hypercube sampling*, which ensures good coverage of prior parameter space (O'Hagan 2006). And for each parameter vector, we store the model output at multiple times. We make all combinations of parameter vectors (n1=9) and times

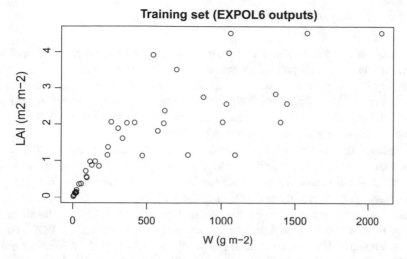

Fig. 14.2 Training set for GP-emulation of EXPOL6: Outputs from 45 input vectors

(nx = 5), giving us a total of n = 45 different input vectors with corresponding EXPOL6s outputs.

```
set.seed(13)
nl          <- 9 ; lhs6 <- lhs::randomLHS( n=nl, k=nb )
sample.b  <- t(sapply( 1:nl, function(i){bmin6 + lhs6[i,]*(bmax6-bmin6)} ))
sample.x  <- c(20,40,60,80,100) / 100 ; nx <- length(sample.x)
sample.xb <- NULL
for (x in sample.x) {sample.xb <- rbind( sample.xb, cbind(x,sample.b) ) }
```

We now run the model for each of the 45 input vectors to create the output vectors (*W*, *LAI*), which are shown in Fig. 14.2. This completes our training data set.

14.4.2 Calibration of the Emulator

We assume values for the hyperparameters of the covariance function: Vy = 3 and phi = 0.2. And we choose a very wide prior for our eight regression parameters which are: the intercept, the slope for the single covariate (time), and the six slopes for the regular parameters of EXPOL6s. Then we calibrate the (regression part of the) GP as we did above for the other examples. That gives us the following values for the posterior means and standard deviations of the eight parameters.

```
> mb_y = ( -7.376 3.013 0.6362 1.784 1.22 1.924 1.936 -0.4282 )
> Standard deviations:  55.05 1.046 52.49 1.872 0.9928 3.09 1.271 1.312
```

14.4.3 Testing the Emulator

Now we want to see how our emulator performs for prediction. We begin with checking the in-sample prediction, where we look at the emulator's interpolation capability for the parameter vectors that were present in the training set of inputs. Remember that the training was only for five times, so what we check here is how the emulator performs for those same parameter vectors at other times. The results of the in-sample prediction are shown in Fig. 14.3.

The lines connect the dots, so the emulator gives the correct results for the training inputs, but we see that uncertainty at intermediate times quickly becomes large. Now let's check the emulator for six parameter vectors randomly drawn from the prior distribution, and not present in the training data. We expect that the emulator's mean function will not correspond to the true values from EXPOL6. The results are shown in Fig. 14.4. We see that the emulator is always close to the actual EXPOL6 curves, but the emulator does not reproduce the pronounced S-shaped LAI-growth curves that EXPOL6 produces with the second and fifth parameter vector.

We test emulator output more thoroughly for t = 0.1, 0.4, 0.7 and 1.0. For each of those times, we compare model and emulator output for 1000 different parameter vectors. Note that usually we shall not be in a position to run such extensive tests: emulators are made when the original model is very slow to run. The results are shown in Fig. 14.5. We see that the emulator differs most from the original model for t=0.1 and also performs poorly for t=1.0. These are times where the EXPOL6 model

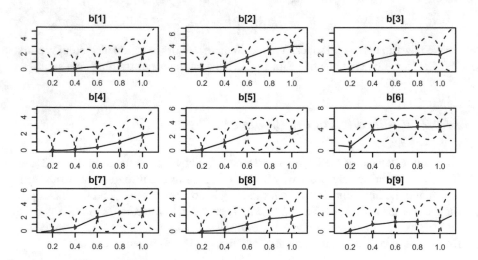

Fig. 14.3 In-sample emulation of EXPOL6: interpolation of LAI $(m^2 m^{-2})$ vs. scaled time for the 9 parameter vectors that were used in GP calibration. Blue dots are the training points, i.e. the LAI-output values from EXPOL6

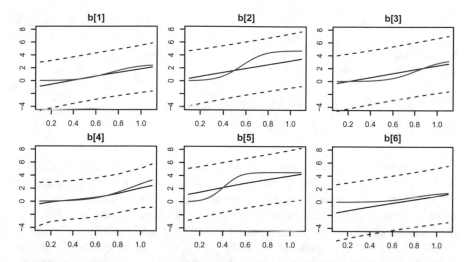

Fig. 14.4 Out-of-sample emulation of EXPOL6: interpolation of LAI (m^2m^{-2}) for 6 parameter vectors that were not used in GP calibration. Blue line: EXPOL6. Black line and dashed lines: emulator

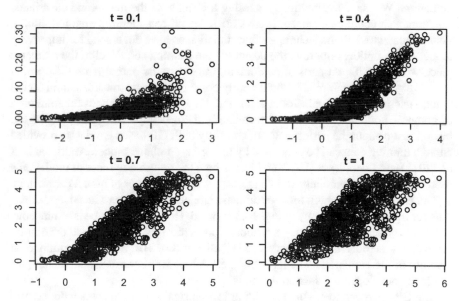

Fig. 14.5 Out-of-sample emulation of EXPOL6: estimation of LAI (m^2m^{-2}) at 4 times for 1000 parameter vectors that were not used in GP calibration

behaves most non-linearly: the time series produced by EXPOL6 for the LAI are exponential in the beginning, then become linear and end asymptotically. Note for example that EXPOL6 can never produce negative outputs, but the emulator does, especially for the earliest times.

14.5 Comments on Emulation

Our two examples of model emulation in this chapter have been simplistic in that we assumed that the covariance function hyperparameters (such as the correlation length constant ϕ in the exponential covariance function) were known. Obviously, these hyperparameters should be estimated rather than guessed, and we discuss the impact of hyperparameter uncertainty in the chapter on spatial modelling (Chap. 22). geoR can estimate hyperparameters but the package cannot be used for emulating functions with more than two parameters. More flexible R-packages for estimating GP-parameters are BACCO and DiceKriging (Roustant et al. 2012).

Besides that, you will probably have many more ideas on how to improve our emulators. We made assumptions regarding the shape of the mean- and covariance functions that could be changed. It pays off to spend considerable time in building your emulator and all the techniques that statisticians have developed for improving model fit are options: improve the design of the training set, develop the emulator piece-wise for different parts of its domain, transform the output from the original model (e.g. a log-transformation for the early phase or a logistic transformation for the whole function domain), etc. Tokmakian et al. (2012) discuss methods for emulation of complex nonlinear models, and fortunately their examples suggest that we can be optimistic about the performance of carefully built GP-based emulators provided enough training data can be generated by the to-be-emulated model. Andrianakis et al. (2015) and Loeppky et al. (2009) discuss how the required number of model runs used for building an emulator should increase with the number of model parameters.

Our emulators were built for deterministic models, so they reproduced model outputs for training data exactly, as Fig. 14.3 showed. However, it can be advantageous to add a so-called *nugget* term—a random noise term $\epsilon \sim N[0, \sigma_\epsilon^2]$—to the emulator, which was found by Andrianakis and Challenor (2012) to improve the numerical stability of the emulation. The term 'nugget' comes from geostatistical techniques originally developed for the mining industry.

The GP-approach to emulator building is empirical and not mechanistic. Hybrid approaches that combine statistical (including machine learning) and mechanistic approaches to emulator building have been studied by Carbajal et al. (2017). It may also be useful to explore ideas from the field of *systems identification* which aims for dynamic surrogate models. An example is the work by Young (1998) who used transfer functions to derive low-dimensional surrogate models that replaced complex climate simulators with systems of just three differential equations which—in stark contrast to GP-emulators—remained interpretable in terms of real world processes.

Exercise

1. GP with small correlation length. Use the geoR-code of this chapter to show that a GP with a linear trend-term and a very small correlation length (ϕ) is equivalent to linear regression as far as estimation of the regression parameters is concerned.

Chapter 15
Graphical Modelling (GM)

A *graphical model* (GM), also called a *probabilistic network*, is a representation of a joint probability distribution. A GM has two parts: (1) a graph with nodes connected by edges, (2) information about the nodes. So the graph is just the visible part of the model. GMs do not represent a new kind of statistical model, they are just helpful tools for analysing joint probability distributions. Every distribution can be represented by a GM, so whatever your research problem or modelling method is, you can choose to use a GM to organize your thinking. And you can choose from multiple kinds of GM as we show in Fig. 15.1. The two most prominent types are *Bayesian networks* (BN, which use directed graphs so that the edges are arrows) and *Markov Random Fields* (MRF, which use undirected graphs). Both classes can represent continuous as well as discrete probability distributions. The different types of GM all come with their own apparatus for designing, interpreting and updating the distributions that they represent. The graphs play an important role in that they indicate ways of decomposing the distribution into conditionally independent parts. We leave this very short general introduction to GM here to concentrate on *Gaussian Bayesian Networks* (GBNs), which can showcase most of the advantages that GM bring. But we shall finish this chapter with some general comments on GM.

15.1 Gaussian Bayesian Networks (GBN)

A *GBN* is a multivariate Gaussian probability distribution, but one for which we have taken the trouble of providing a nice depiction in the form of a *Directed Acyclic Graph (DAG)*. Every multivariate Gaussian is fully specified by its mean vector and covariance matrix but these are not shown in the graph. Instead, the graph shows how the distribution can most efficiently be factorised. GBNs are a subset of *Bayesian Networks* (BN, also called *Bayesian Belief Networks*, BBN). BNs can represent every

© Springer Nature Switzerland AG 2020
M. van Oijen, *Bayesian Compendium*,
https://doi.org/10.1007/978-3-030-55897-0_15

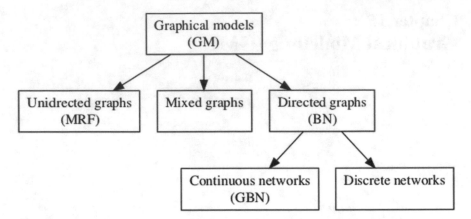

Fig. 15.1 Simplified scheme of graphical modelling categories with only the major types shown

Fig. 15.2 The simplest
Directed Acyclic Graph
(DAG)

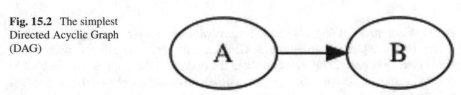

joint probability distribution, including ones for which some or all of the marginal distributions are not continuous as in a GBN, but discrete.

The DAG for the simplest GBN has two nodes linked by one edge (Fig. 15.2). The nodes represent the variables we are uncertain about, and the edge shows how we propose to decompose their joint distribution.

For such a simple GBN, factorisation of the joint probability distribution, $p[A, B]$, is trivial. The arrow in the figure goes from A to B, which indicates that $p[A, B]$ is to be factorised as $p[A]p[B|A]$, but the opposite factorisation, $p[B]p[A|B]$, is mathematically equivalent and may be just as simple. It is only when our networks are more complicated that some factorisations may be simpler (have fewer edges) than others. For such networks the ordering of nodes we show in the graph, using the arrows, becomes important. If we let the arrows represent causality, i.e. running from causes to effects, we generally produce the simplest DAG.

15.1.1 Conditional Independence

Say we have a network with three nodes A, B and C. Then we can always write the joint pdf as $p[A, B, C] = p[A]p[B|A]p[C|A, B]$, or the same equation with any other permutation of A, B and C. The corresponding graph would have three arrows: from A to both B and C, and from B to C. But what if we knew that the relationship between A and C was fully mediated by B? For example, A = resources, B = growth

Fig. 15.3 C is conditionally independent of A

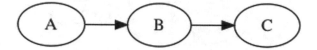

rate, C = yield, where we assume that any correlation between resources and yield is fully due to resources stimulating growth and that in turn leading to high yield. In such a case we say that C is conditionally independent of A. Our DAG would then have just two arrows (Fig. 15.3).

This DAG represents our belief that $p[A, B, C]$ can be factorised simply as $p[A]p[B|A]p[C|B]$. The factorisation implied by the graph may be the most natural (it is a chain of cause and effect) but other factorisations are possible that represent exactly the same joint probability distribution. By the way, this three-node probability chain is the simplest example of a *hierarchical model*.

15.2 Three Mathematically Equivalent Specifications of a Multivariate Gaussian

GBNs represent multivariate Gaussians graphically, but how do we specify the underlying probability distribution quantitatively? The most common way is to give the mean vector and the covariance matrix (or its inverse, the *precision matrix*). The diagonal of the covariance matrix gives the *unconditional variances* of the nodes, which quantify our uncertainty about the node-values when the values of the other nodes are not known yet. Our knowledge about each node is then not yet constrained, or 'conditioned', by knowledge of the other nodes.

But sometimes a quite different specification is more useful where we say how much uncertainty we still have about each node after the values of all other nodes have in fact become known. That is done by giving the vector of *conditional variances*. That vector of course tells us nothing about the relationships between the nodes but if we complement it by also giving the regression coefficients for each pair of nodes and further specify the mean vector too, then we have again a full specification of the multivariate Gaussian underlying our GBN.

Say that we have that DAG for a GBN with n nodes and r edges. We have seen that there are three different ways of specifying the multivariate Gaussian distribution for this GBN:

1. $n \times 1$ mean vector + $n \times n$ covariance matrix,
2. $n \times 1$ mean vector + $n \times n$ precision matrix,
3. $n \times 1$ mean vector + $n \times 1$ vector of conditional variances + $r \times 1$ vector of regression coefficients.

All three methods have their uses and will appear in this chapter. But note that only the third method actually makes use of the information in the DAG, and it is also often the easiest to specify as we shall see. And the vector-information that it

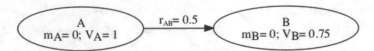

Fig. 15.4 DAG with means and conditional variances specified in the node-ellipses and edge labelled with the regression coefficient

provides could be shown in the DAG itself, which would make the DAG fully self-explanatory. All we need to do for that is adding the mean and conditional variance to each node, and labelling each edge with the regression coefficient. See Fig. 15.4 for an example.

A convenient property of GBNs is that any two nodes for which the regression coefficient is zero are conditionally independent. That tells us that we do not need to draw an arrow between those nodes. In fact, the main use of the graph is to show which nodes are conditionally independent: absence of an arrow tells us more than the presence of one.

15.2.1 Switching Between the Three Different Specifications of the Multivariate Gaussian

It is easy to switch between the three ways of specifying the multivariate Gaussian. Conversion between (1) and (2) is conceptually the easiest: the covariance and precision matrices are each other's inverse. So, if the two matrices are denoted as Σ resp. W, then $W = \Sigma^{-1}$, which in R-code would be written as 'solve(Sigma)'.

The conversion from (3) to (2) can be done using the algorithm of Shachter and Kenley (1989), which we code in R as follows:

```
precMatrix <- function( nodes, Vcond, R ){
  n <- length(nodes) ; W <- 1 / Vcond[1]
  for(k in 2:n){
    rk        <- R[ k, 1:(k-1) ]
    W_top     <- cbind( W * Vcond[k] + rk %*% t(rk), -rk ) / Vcond[k]
    W_bottom  <- cbind(                      -t(rk),   1 ) / Vcond[k]
    W         <- rbind( W_top, W_bottom ) }
  rownames(W) <- colnames(W) <- nodes
  return(W) }
```

This function has three arguments: 'nodes', 'Vcond' and 'R'. The first argument is the vector of node-names. The second is an equally long vector of conditional variances. And the third is a matrix R which has zeroes everywhere except for the elements R_{ij} that correspond to an edge from node j to node i in the DAG. Let's apply this to the example DAG of Fig. 15.4. The following three lines of code give the three different specifications. The code does not specify the mean vector, but it is the same for all three methods, and we choose $\mu = (0, 0)$.

```
nodes <- c("A","B") ; Vcond <- c(1,0.75) ; R <- matrix( c(0,0.5,0,0), nrow=2)
W       <- precMatrix( nodes, Vcond, R )
Sigma <- solve(W)
```

So the distribution implied by Fig. 15.4 has the following mean vector, covariance matrix and precision matrix:

GAUSSIAN derived from DAG:

$$\mu = \begin{bmatrix} 0 \\ 0 \end{bmatrix}; \quad \Sigma = \begin{bmatrix} 1 & 0.5 \\ 0.5 & 1 \end{bmatrix}; \quad W = \begin{bmatrix} 1.333 & -0.667 \\ -0.667 & 1.333 \end{bmatrix}. \tag{15.1}$$

This shows that the conditional variance of B that was specified in the DAG ($V_B = 0.75$) is less than its unconditional variance in the covariance matrix ($\Sigma[2, 2] = 1$) because knowledge of A reduces uncertainty about B, thanks to the non-zero regression coefficient ($R[2, 1] = 0.5$). In contrast, the conditional variance of A in the DAG ($V_A = 1$) is the same as its unconditional variance ($\Sigma[1, 1] = 1$) because we use the factorisation of the joint probability distribution where A is a parent of B and not the other way around. Note that this lack of symmetry between A and B is in no way inherent or visible in our mean vector and covariance matrix. It is simply the result of us specifying an ordering by means of the DAG, which defines how the regression coefficients are calculated—in this case for B as a function of A and not vice versa.

Castillo et al. (2008) provided an algorithm for the conversion from method (2) to (3) that we implement here as follows:

```
VcondR <- function( W ) {
  n   <- dim(W)[1] ; Vcond <- rep( NA, n )
  Wk <- W          ; R     <- matrix( 0, nrow=n, ncol=n )
  for(k in n:2){
    ik        <- 1 : (k-1)
    Vcond[k] <- 1 / Wk[ k, k]
    R[k,ik]  <-      -Wk[ k,ik] * Vcond[k]
    Wk       <-       Wk[ik,ik] - as.matrix(R[k,ik]) %*% R[k,ik] / Vcond[k] }
  Vcond[1] <- 1 / Wk[1,1]
  return( list( Vcond=Vcond, R=zapsmall(R) ) ) }
```

Above we derived the precision matrix from the DAG of Fig. 15.4. Let's verify that our new function gives us back the conditional variances and regression coefficients when we apply it to that precision matrix. So we run the command:

```
VcondR(W)
```

and that indeed retrieves the conditional variances and regression coefficients that we started out from:

COND. VARS and REGR. COEFF. derived from PRECISION MATRIX:

$$\text{Vcond} = \begin{bmatrix} 1 \\ 0.75 \end{bmatrix}; \quad R = \begin{bmatrix} 0 & 0 \\ 0.5 & 0 \end{bmatrix}. \tag{15.2}$$

We now have the tools available for our study of GBNs in the remainder of this chapter.

15.3   The Simplest DAG Is the Causal One!

A good DAG that follows causal reasoning (with edges going from causes to consequences) will have fewer edges than non-causal DAGs for the same joint probability
distribution. Let's study this with an example. We specify a causal DAG for the
impacts of rain and irrigation on soil water based on the following code chunk:

```
nodes <- c("Rain","Irrigation","SoilWater")
m     <- c( 2    , 0.5         , 1        )
Vcond <- c( 0.2  , 0.05        , 0.1      )
R     <- matrix( rep(0,9), nrow=3) ; R[3,1] <- 0.9 ; R[3,2] <- 1
```

The causal DAG is drawn on the left in Fig. 15.5. It has three nodes but only
two edges (non-zero values in the R matrix). If we wanted to make life difficult for
ourselves, we could: (1) Derive the covariance matrix for this DAG, (2) Reorder the
mean vector to ("SoilWater", "Irrigation", "Rain") and reorder the covariance matrix
accordingly, (3) Derive the conditional variances and regression coefficients for this
new ordering. The non-causal DAG that follows from that exercise is shown in the
same figure on the right.

The non-causal DAG has three edges, which is the maximum that a three-node
DAG can have. So it is more complicated than the two-edges causal DAG, despite the
fact that both DAGs represent exactly the same joint probability distribution! Causal
thinking helps us keep our DAGs simple, which makes them much easier to work
with.

15.4   Sampling from a GBN and Bayesian Updating

Because a GBN is a probability distribution, we can sample from it. This can be
done one node at a time, following the factorisation implied by the structure of the
DAG, but we can also do this using standard algorithms for sampling vectors from
the whole multivariate Gaussian in one go. The first method—which only involves
conditionally *univariate* Gaussians—will be computationally more efficient when
the network has many nodes because it avoids the inversion of the covariance matrix.

15.4.1   Updating a GBN When Information About Nodes
         Becomes Available

After we have specified our network, new information may become available. For
example, the value of one or more nodes may become known through measurement

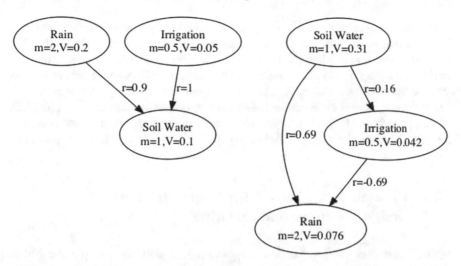

Fig. 15.5 Two DAGS for the same probability distribution. The causal DAG on the left is smaller than the non-causal one on the right

or observation. When such new information becomes available, in any BBN, not just GBNs, then that information propagates to the remaining nodes. If the *evidential nodes* have no *parents*, just *children*, then updating the network is just a matter of basic forward propagation of information following the direction of the arrows to all the *descendants* of the evidential nodes. The situation is more complicated if the evidential nodes themselves are children of other nodes. But then Bayes' Theorem can be used to update the distribution.

In the case of GBNs, Bayesian updating of the network is very easy: there is an analytical solution to the problem of finding the 'posterior network', so we need not worry about how to implement Bayes' Theorem. That analytical solution is implemented in R here. The algorithm operates on the covariance matrix of the GBN so for this task we need to provide the covariance specification of the GBN. [There are computationally more efficient ways that take advantage of the network structure but those take more steps and are not needed for the simple examples of this chapter.] If all we have is the specification in terms of the conditional variances and regression coefficients, then we must first convert that specification into the covariance matrix (via the precision matrix) with the code shown above. Here is code for the posterior mean and covariance matrix of a multivariate Gaussian distribution after we learn the values y for one or more of the nodes.

```
GaussCond <- function( mz, Sz, y ) {
  i <- 1 : ( length(mz) - length(y) )
  m <- mz[i]   + Sz[i,-i] %*% solve(Sz[-i,-i]) %*% (y-mz[-i])
  S <- Sz[i,i] - Sz[i,-i] %*% solve(Sz[-i,-i]) %*% Sz[-i,i]
  return( list( m=m, S=S ) ) }
```

In this implementation, the vector y (with length n_y) represents the nodes that become known from, for example, measurement. The code assumes that the measurements correspond to (and in a way replace) the last n_y values of the mean vector. The function returns the new mean vector and covariance matrix for the smaller network that remains when the known nodes are removed and the others updated. Let's use it for a very simple example, using the same DAG as above, where we now assume that we learn that B=2. The function thus returns the posterior mean and variance of A. This is coded as GaussCond(c(0,0), Sigma, 2) and tells us that $p[A|B] = N[1, 0.75]$. Compare the posterior conditional mean and variance for A with the numbers in the DAG of Fig. 15.2. Are the results as you expected?

15.5   Example I: A 4-Node GBN Demonstrating DAG Design, Sampling and Updating

We shall now illustrate some uses of graphical models with the example of a 4-node GBN. We start from the mean + covariance matrix specification:

```
nodes <- c( "A", "B", "C", "D" )
m     <- c( 3 , 4 , 9 ,  14 )
S     <- matrix( c(4 ,   4,   8, 12,
                   4 ,   5,   8, 13,
                   8 ,   8,  20, 28,
                  12, 13,  28, 42), nrow=4 )
```

We now calculate the conditional variances and the regression coefficients to prepare us for plotting the DAG:

```
VR <- VcondR( solve(S) ) ; Vcond <- signif(VR$Vcond,2) ; R <- VR$R
```

When we inspect R, we see that there are only four non-zero regression coefficients. So we need only four edges in our DAG. It would have been difficult to conclude directly from the covariance matrix that there are so many conditional independencies in this GBN! (Fig. 15.6).

If we want to assess the uncertainty represented by this 4-node GBN we should sample from it. And because we know the mean vector m and covariance matrix S, we could sample with the R-command rmvnorm(100,m,S). Because our GBN does not have many nodes, that will not be computationally demanding. However, let us instead apply the method that is more suitable for large GBNs, namely sequential sampling, one node at a time. To do that efficiently, we follow the ordering of the nodes in the DAG and make use of the regression coefficients, as in the following code:

```
n <- 100 ; sampleSeq <- sapply( LETTERS[1:4], function(x) NULL )
sampleSeq$A <- m[1] + rnorm( n, 0, sqrt(Vcond[1]) )
sampleSeq$B <- m[2] + rnorm( n, 0, sqrt(Vcond[2]) ) +
                      R[2,1] * (sampleSeq$A - m[1])
sampleSeq$C <- m[3] + rnorm( n, 0, sqrt(Vcond[3]) ) +
                      R[3,1] * (sampleSeq$A - m[1])
sampleSeq$D <- m[4] + rnorm( n, 0, sqrt(Vcond[4]) ) +
                      R[4,2] * (sampleSeq$B - m[2]) +
                      R[4,3] * (sampleSeq$C - m[3])
lapply( sampleSeq, mean )
```

Fig. 15.6 DAG with four
nodes

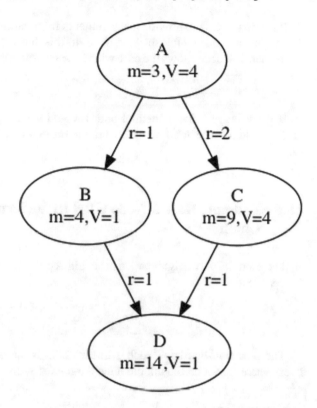

You can verify for yourself that this produces a representative sample from the
multivariate Gaussian of the GBN.

Now let's assume that the first node, A, becomes known through measurement.
Say A = 7. What does that mean for the remaining nodes? Well, we defined the function GaussCond <- function(mz, Sz, y) that gives us the posterior
multivariate Gaussian when one or more nodes becomes known. Unfortunately the
function definition requires the known nodes to be the *last* ones in the specification,
and our node A is the first. So we need to reorder the mean vector and covariance
matrix by putting A at the end before applying the function.

```
i2 <- c(2:4,1) ; nodes2 <- nodes[i2] ; m2 <- m[i2] ; S2 <- S[i2,i2]
GaussCond(m2,S2,7)
> $m
>          [,1]
> [1,]     8
> [2,]    17
> [3,]    26
>
> $S
>        [,1] [,2] [,3]
> [1,]    1    0    1
> [2,]    0    4    4
> [3,]    1    4    6
```

It is also possible to provide measurements from multiple nodes at the same time. The following code chunk shows how to do this, but we leave running the code to the reader. The code shows the case where the measurements are {A=7,B=8,C=17}.

```
i3 <- c(4,1:3) ; nodes3 <- nodes[i3] ; m3 <- m[i3] ; S3 <- S[i3,i3]
GaussCond(m3,S3,c(7,8,17))
```

If the values for the evidential nodes would have been different, the posterior mean would have been different too, but the posterior variance would have been the same.

15.6 Example II: A 5-Node GBN in the form of a Linear Chain

In this example, only neighbours in the chain A->B->C->D->E are not conditionally independent.

```
n <- 5                      ; nodes <- LETTERS[1:n]
m <- rep(0,n)               ; Vcond <- rep(1,n)
R <- matrix(0,nrow=n,ncol=n) ; for(j in 2:n){ R[j,j-1] <- 0.5}
```

The precision matrix of such a linear chain is characterised by having zero's everywhere except on the middle three diagonals:

```
precMatrix( nodes, Vcond, R )
>       A      B      C      D      E
> A  1.25 -0.50  0.00  0.00  0.0
> B -0.50  1.25 -0.50  0.00  0.0
> C  0.00 -0.50  1.25 -0.50  0.0
> D  0.00  0.00 -0.50  1.25 -0.5
> E  0.00  0.00  0.00 -0.50  1.0
```

15.7 Examples III & IV: All Relationships in a GBN are Linear

The relationships in a GBN are linear: all children depend linearly on their parents. The slopes of the linear relationships are given by the regression coefficients R_{ij}. The strengths of the relationships can be assessed by inspecting the covariance matrix. Some nodes may be so closely correlated that the relationship is effectively a deterministic linear function as we show in the immediately following Example III. That will be followed by Example IV demonstrating how GBNs can represent multivariate stochastic linear relationships.

Fig. 15.7 Sampling from a
multivariate Gaussian that
mimics a near-deterministic
linear relationship

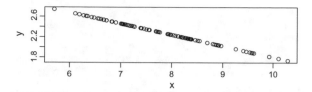

15.7.1 Example III: A GBN Representing Univariate Linear Dependency

To show that relationships in a GBN can be nearly deterministic, let's construct a
2-node GBN with a near-deterministic relationship, given as $y = 4 - 0.22x + \sigma_\epsilon$,
with $\mu_x \sim N(8, 1)$ and $\sigma_\epsilon^2 = 10^{-6}$. We encode this as a bivariate Gaussian in the
following way:

```
a  <- 4 ; b  <- -0.22
mx <- 8 ; Vx <- 1 ; my <- a + b * mx ; m <- c( mx, my ) ; Ve <- 1.e-6
S  <- matrix( c(  Vx, b  *Vx, b*Vx, b^2*Vx + Ve), nrow=2 )
```

Let's sample from this multivariate Gaussian to show that we have indeed represented the intended near-deterministic dependency of y on x (Fig. 15.7):

Now let's represent this relationship as a GBN. For that we need the conditional
variances of x and y and the regression coefficient. We calculate them in the usual
way from the covariance matrix:

```
VR <- VcondR( solve(S) ) ; Vcond <- signif(VR$Vcond,2) ; R <- VR$R
```

We shall not plot the DAG, but it would simply consist of two nodes $x \rightarrow y$, with
mean values (8, 2.24), conditional variances $(1, 10^{-6})$, and with regression coefficient
-0.22. Note that in this simple case, we could have inferred the conditional variances
and the regression coefficient directly from our premises ($y = 4 - 0.22x + \sigma_\epsilon, \sigma_x^2 = 1, \sigma_\epsilon^2 = 10^{-6}$). But the main point of this section is to show that GBNs, though
limited in that all conditional probabilities must be Gaussian, can include (nearly)
deterministic influences of some nodes on others.

15.7.2 Example IV: A GBN Representing Multivariate Stochastic Linear Relations

This time we show that GBNs can also represent a linear sum of terms. We'll make
the relationship more strongly stochastic this time, so not nearly deterministic. Say
$z = 2 + 3x + 5y$, with $\mu_x = 10, \mu_y = 1$ and $V_x = V_y = 1$. We assume no linear
relationship between x and y.

```
a   <-   2 ; b1 <- 3 ; b2 <- 5
mx <- 10 ; Vx <- 1 ; my <-  1 ; Vy <- 1 ; mz <- a + b1 * mx + b2 * my
m   <- c( mx, my, mz ) ; Ve <- 10
S   <- matrix( c(    Vx,      0, b1  *Vx,
                      0,     Vy,         b2*  Vy,
                   b1*Vx, b2*Vy, b1^2*Vx + b2^2*Vy + Ve ), nrow=3 )
```

So we now constructed the mean vector and covariance matrix of the multivariate Gaussian representing $z = 2 + 3x + 5y$ with the given variances. We sample from it and (for once) use R's 'lm'-function to confirm that we have produced data from the desired relationship.

```
n              <- 1e5
samplexyz <- rmvnorm( n, m, S )
samplex <- samplexyz[,1]
sampley <- samplexyz[,2]
samplez <- samplexyz[,3]
lm( samplez ~ samplex + sampley )
>
> Call:
> lm(formula = samplez ~ samplex + sampley)
>
> Coefficients:
> (Intercept)       samplex        sampley
>        2.148         2.987          4.990
```

15.8   Example V: GBNs can do Geostatistical Interpolation

In our final example of graphical modelling, the nodes now represent n different points in space. Our prior uncertainty about the node values will be represented by a multivariate Gaussian with covariances that depend on the distances between points. We shall use a negative exponential covariance function with a given correlation length ϕ, i.e. $Cov(i, j) = exp(-distance(i, j)/\phi)$. We choose the mean vector to be equal to 2 everywhere, and the variances equal to 1.47. We assume that there are $n = 4$ points of interest, named "s1" to "s4" and located at coordinates (0,3), (4,0), (4,3) and (0,0) respectively. We assume that $\phi = 1.9115$. That defines our multivariate distribution as follows:

```
n     <- 4 ; nodes <- paste0( "s", 1:4 ) ; m <- rep( 2, n ) ; phi <- 1.9115
dist <- matrix( c( 0, 5, 4, 3,
                   5, 0, 3, 4,
                   4, 3, 0, 5,
                   3, 4, 5, 0 ), nrow=4 )
S     <- 1.47 * exp(-dist/phi)
```

In a typical geostatistics problem, we get measurement values for all nodes except one. The problem is then to infer the posterior distribution for that unmeasured node. Once we know how to do that, we can move our unmeasured location to any location in our region of interest, get our estimates for that point, and so on. That way we can build up a map for the whole region when only a finite number of points have been measured. That method is called *kriging*. And in this chapter, we shall be using a

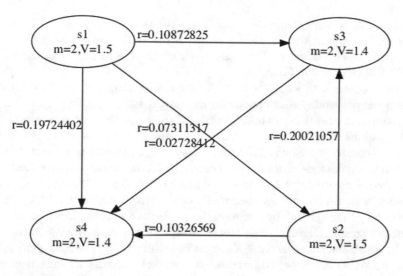

Fig. 15.8 DAG for kriging

graphical model for that. So let's make a GBN, with its DAG, for our spatial network of nodes (Fig. 15.8).

```
VR <- VcondR( solve(S) ) ; Vcond <- signif(VR$Vcond,2) ; R <- VR$R
```

Now let's assume that we measure values at three of the locations: {s1=0.7, s2=3.1, s3=2.2}. So we want to estimate the probability distribution for s4. Let's calculate the posterior distribution for s4 in the same way as we have done before in this chapter.

```
i2 <- c(4,1:3) ; nodes2 < nodes[i2] ; m2 <- m[i2] ; S2 <- S[i2,i2]
y  <- c(0.7,3.1,2.2) ; GaussCond(m2,S2,y)
> $m
>          [,1]
> [1,] 1.862632
>
> $S
>          [,1]
> [1,] 1.387985
```

So the kriging prediction for point s4 was a mean value of 1.86 with a kriging variance of 1.39. This was of course a very simple example of geostatistics, where we assumed that a priori the mean value everywhere was 2, independent of the coordinates. More usually, in geostatistics, we first fit a trend-surface to the data, and model the residuals using the multivariate Gaussian. For that approach, see Chap. 22.

15.9 Comments on Graphical Modelling

Graphical modelling is a widely applicable tool for formulating and visualizing statistical models. All statistical models can be formulated as a joint probability distribution that we depict in a graph, but only some have a tradition of doing so.

These include *structural equation modelling*, *path analysis*, *Hidden Markov Models* (HMM), and *hierarchical modelling* which is the subject of the next chapter. A graphical model that presents the general structure of *Bayesian Decision Theory* (BDT) will be shown in Chap. 17.

Our examples in this chapter were all GBNs, which made it easy to specify the conditional probability distributions for each node. But Bayesian networks are in fact more often used to depict quite irregular discrete distributions, where each node comes with its own 'Conditional Probability Table' (Hojsgaard et al. 2012). This is partly because commercial software for GM design and calibration can currently not handle continuous distributions. Discrete Bayesian networks have frequently been used in environmental management and have shown their use for facilitating stakeholder involvement (Gonzalez-Redin et al. 2016; Smith et al. 2012). People can co-design the network by discussing the variables and influences that the graph needs to represent. The discussions focus on representing causality which keeps the graphs sparse and understandable. Contrast this with our first example above which showed that the causality-agnostic covariance matrix looked much more complicated than the graph we derived from it. The graph revealed many unsuspected conditional independencies.

Exercises

1. Bayes' Theorem as a DAG. If you wanted to show Bayes' Theorem as a two-node DAG, would you choose $[\theta] \rightarrow [y]$ or $[y] \rightarrow [\theta]$?

2. Covariance matrices (after MacKay). Which of the following four matrices could be the covariance matrix of a GBN with three edges: $y1 \rightarrow y2$, $y1 \rightarrow y3$, $y2 \rightarrow y3$?

```
M1 = matrix( c(9,   3, 1,   3,   9,  3, 1,   3, 9), nrow=3 )
M2 = matrix( c(8,  -3, 1,  -3,   9, -3, 1,  -3, 8), nrow=3 )
M3 = matrix( c(9,   3, 0,   3,   9,  3, 0,   3, 9), nrow=3 )
M4 = matrix( c(9,  -3, 0,  -3,  10, -3, 0,  -3, 9), nrow=3 )
```

Chapter 16
Bayesian Hierarchical Modelling (BHM)

In the previous chapters, our statistical procedure was very simple: define a prior probability distribution for the parameters $p[\theta]$ and a likelihood function $L[\theta] = p[y|\theta]$, and that was it. Bayes' theorem then told us what the posterior distribution would be once we received the data: $p[\theta|y] \propto p[\theta]L[\theta]$. The prior for the parameter vector was always a fully specified distribution, e.g. the product of known univariate Gaussians. In hierarchical modelling, we do not specify the prior that directly. Instead we make the prior distribution depend on other parameters, which we call *hyperparameters*. Here is a table of the differences:

	Non-hierarchical	Hierarchical	
Likelihood	$p[y|\theta]$	$p[y|\theta]$	
Prior	$p[\theta]$	$p[\theta|\xi]$	(16.1)
Hyperprior	–	$p[\xi]$	

Strictly, hierarchical models are nothing new, because they can always be collapsed into non-hierarchical models. You do that by combining the parameters and hyperparameters in a new longer parameter vector $\omega = (\theta, \xi)$ and finding the 'flattened', non-hierarchical prior distribution by multiplication: $p[\omega] = p[\theta|\xi]p[\xi]$.

When we carry out Bayesian calibration of the parameters and hyperparameters in a hierarchical model, statisticians refer to the approach as Bayesian Hierarchical Modelling (Cressie and Wikle 2011; BHM). In BHM, we must specify a prior for the hyperparameters, $p[\xi]$, which then automatically implies a prior for the regular parameters, $p[\theta|\xi]$. But the likelihood function does not change in this hierarchical setup, because model output $f(x, \theta)$ only depends on the regular model parameters, not the hyperparameters. So, in BHM, the posterior distribution is defined as:

$$p[\theta, \xi|y] \propto p[\xi]\, p[\theta|\xi]\, L[\theta] \tag{16.2}$$

© Springer Nature Switzerland AG 2020
M. van Oijen, *Bayesian Compendium*,
https://doi.org/10.1007/978-3-030-55897-0_16

Note that the use of BHM implies greater data needs. For example, if the hyperparameters in the BHM stand for spatial variability, then calibration cannot be effective without data from dispersed sites.

16.1 Why Hierarchical Modelling?

So when would this decomposition of the prior be useful? Well, it turns out that it is often very convenient to think hierarchically. The parameters θ can represent properties of individuals, and the hyperparameters ξ properties of the population(s) to which the individuals belong (Cressie et al. 2009; Ogle 2009; Ogle and Barber 2008). For example, modellers need not assume that all trees in a region have the same wood density, but can allow for spatial variation represented by two hyperparameters: mean wood density and its standard deviation. So they would represent the wood density of n individual trees as $\theta = (\theta_1, ..., \theta_n)$ and the mean and variance of those wood densities as $\xi = (\mu, \sigma^2)$.

In disciplines such as ecology, hierarchical modelling is rapidly becoming the standard approach. Ecosystem modellers are giving increased attention to how various plant and animal traits vary with environmental conditions (Simpson et al. 2016). Hierarchical modelling allows for great flexibility in such cases: each parameter can range from being completely generic (zero population variance) to being completely site- or condition-specific (large population variance). A good introduction to hierarchical modelling in the context of vegetation research was provided by Dietze et al. (2008). They quantified variability in parameters for tree allometry (power law relationships for height vs. crown area) both within and between species in southeastern US forests, and showed how even species for which few data were available could be fit reliably within the hierarchical model.

But the scope for using BHM goes well beyond representing spatial or taxonomic differences: the many different sources of error mentioned in this book can all be represented explicitly, with their own hyperparameters, in a hierarchical statistical model (Cressie et al. 2009). For example, BHM facilitates the use of different data sets in one Bayesian calibration whereby differences in measurement precision and accuracy are represented by data set-specific and generic error-parameters (Ogle 2009). Bayesian hierarchical modelling is therefore also the obvious tool for *meta-analysis*, where we want to extract robust conclusions from multiple observational studies that partly, but not completely, share similar measurement methods (DuMouchel and Normand 2000; Ogle et al. 2014; Williams et al. 2018).

16.2 Comparing Non-hierarchical and Hierarchical Models

Let's make it all less abstract by comparing three statistical models for linear regression: two non-hierarchical ones and a hierarchical model. The three models are depicted as graphical models in Fig. 16.1.

Our starting point is as usual Equation (2.4): $p[y|\theta] = p[f(x, \theta) - y = \epsilon_y]$, where we now choose $f(x, \theta) = a + bx$ with $\theta = (a, b)$. We expect data $y = \{y_{ij}\}$ from J different species, with n_j measurements per species, so $y = \{y_{ij}\}$ with $(i = 1 : n_j, \ j = 1 : J)$. We assume that measurement error has a Gaussian distribution $\epsilon_y \sim N[0, \sigma_y^2]$ with known variance σ_y^2. All this is quite standard. Our three models (A, B, C) only differ in their assumptions about the paramaters a and b. In model A, both parameters are generic so they have the same value for all species. In models B and C, the intercepts and slopes vary across species. But only in model C are we going to estimate hyperparameters that represent the intercept- and slope-variation between species. Despite these differences, the three models have, in a general sense, the same likelihood function:

$$p[y|\theta] = \prod_{j=1}^{J} \prod_{i=1}^{n_j} p[y_{ij}|\theta], \tag{16.3}$$

but the likelihood functions for individual data points $p[y_{ij}|\theta]$ do differ:

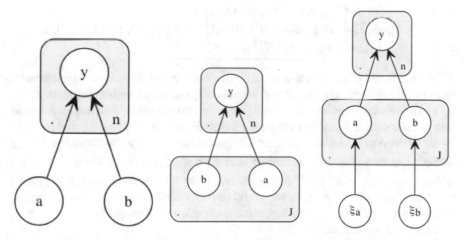

Fig. 16.1 Three statistical models for n observations on J different grassland cultivars. From left to right: Model A is a non-hierarchical model with universal slope ('a') and intercept ('b'); Model B is also non-hierarchical but has cultivar-specific parameters a and b; Model C is a Bayesian Hierarchical Model (BHM) with cultivar-specific parameters and global hyperparameters. We use plate-notation where the 'n' and 'J' in bottom-right corners indicate the multiplicity of data and cultivars

Model	Parameter vector	$p[y_{ij}|\theta]$
A	$\theta = (a, b)$	$N[y_{ij}|a + bx_{ij}, \sigma_y^2]$
B and C	$\theta = (a_1..a_J, b_1..b_J)$	$N[y_{ij}|a_j + b_j x_{ij}, \sigma_y^2]$.

$$(16.4)$$

The priors for parameters (and hyperparameters in the case of model C) also differ. For the last model, we use four hyperparameters to represent the distributions of intercept and slope among the species: $\xi = (\xi_a, \xi_b) = (\mu_a, \sigma_a^2, \mu_b, \sigma_b^2)$.

Model	Parameter vector	Prior
A	$\theta = (a, b)$	$p[\theta] = p[a]p[b]$
B	$\theta = (a_1..a_J, b)$	$p[\theta] = p[a_1]..p[a_J]p[b_1]..p[b_J]$
C	$(\theta, \xi) = (a_1..a_J, b_1..b_J, \mu_a, \sigma_a^2, \mu_b, \sigma_b^2)$	$p[\theta, \xi] = p[\theta|\xi]p[\xi] =$
		$p[a_1|\mu_a, \sigma_a^2] .. p[a_J|\mu_a, \sigma_a^2] \times$
		$p[b_1|\mu_b, \sigma_b^2] .. p[b_J|\mu_b, \sigma_b^2] \times$
		$p[\xi_a] \, p[\xi_b]$.

$$(16.5)$$

Let's choose Gaussian priors for parameters and the two 'mean' hyperparameters, and uniform priors for the two 'variance' hyperparameters (not Gaussians because variances cannot be negative):

Model	Prior
A	$p[\theta] = N[a|500, 10^6] \, N[b|0, 10^2]$
B	$p[\theta] = \prod_{j=1}^{J} N[a_j|500, 10^6]N[b_j|0, 10^2]$
C	$p[\theta, \xi] = N[\mu_a|500, 10^6] \, N[\mu_b|1, 10^6] \, U[\sigma_a^2|0, 10^3] \, U[\sigma_b^2|0, 2] \times$
	$\prod_{j=1}^{J} N[a_j|\mu_a, \sigma_a^2] \, N[b_j|\mu_b, \sigma_b^2]$.

$$(16.6)$$

Models B and C are the most realistic models here, as they allow all parameters (a,b) to vary between species. But if we get data, and the posterior distributions for σ_a^2 and σ_b^2 are such that only very small variances are plausible, then model A would be a good approximation. But we must be careful here. It may be tempting to begin by using the data y to find maximum likelihood estimates for the variances, i.e. $\widehat{\sigma_a^2}$ and $\widehat{\sigma_b^2}$, and then plug those estimates into the prior distributions for model A, i.e. $p[\theta] = p[a] \, p[b] = N[0, \widehat{\sigma_a^2}] \, N[0, \widehat{\sigma_b^2}]$. But if we then calibrate model A against the data using those priors, then we would be using the data twice, and that is illegal. The practice is called *empirical Bayes* and is of course not a proper Bayesian method, and it should be avoided.

Let's reflect some more about the differences between the three models. In model A, the parameters are generic, i.e. they are assumed to have the same value for all species. If that is acceptable, then that would of course be very good, because our model could then be used everywhere without species-specific recalibration. Model B is more modest, it assumes that intercepts and slopes are species-specific. But it makes no assumptions about the variation of those parameters between species. And

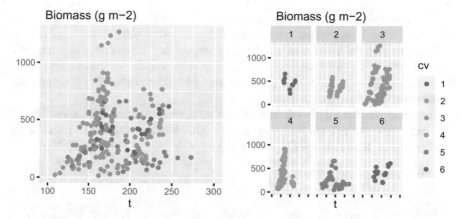

Fig. 16.2 Grassland growth data used for Bayesian calibration of three models. 't' = time (days), 'cv' = cultivar

what we learn about the intercept and slope of one species, does not tell us anything about the value of these parameters for any other species. Model C is much more useful in that respect, because if you calibrate that model, you are also learning about interspecific variation in the parameter values. We can use that information about the hyperparameters (μ_a, σ_a^2) and (μ_b, σ_b^2) when we want to estimate slopes and intercepts for new species. In short, hyperparameters define the prior for what to expect for the regular parameters (a, b) of new species.

Let's now implement the three models. We shall be using real data on the growth of six different grassland varieties. Our data set contains 258 data points (t, y), where t is time in days and y is plant dry matter $(g\, m^{-2})$. The number of data points for the varieties $\{n_j\}$, $j = 1, .., 6$ is $\{10, 42, 80, 76, 36, 14\}$. The data are shown in Fig. 16.2.

16.2.1 Model A: Global Intercept and Slope, Not Hierarchical

To define our models, we use the R-package 'rjags', which allows us to formulate the models in the 'JAGS' language for carrying out Bayesian calibration using MCMC. This goes against the promise that we would use basic R code where possible, but JAGS is convenient (especially for model C below) and you have seen the calibration of simple linear models with our own MCMC-code already in other chapters.

In JAGS, the first model is defined as follows:

```
ModelA <- " model {
   for (i in 1:ny) {
      y[i] ~ dnorm( a + b * t[i], tau.y[i] )
      tau.y[i] <- pow( sigma.y[i], -2 ) }
   a ~ dnorm( 5e2, 1e-6 ) ; b ~ dnorm( 1, 1e-2 ) } "
```

The main thing to notice is that the intercept a and slope b are not indexed, so they are universal parameters. The last line of the code shows the Gaussian priors for the two parameters. Note that JAGS has its own 'dnorm' command which requires specification of the *precision* rather than the *variance* as its second argument, so the small values we chose for the precision imply large prior uncertainty about both parameters.

To actually run the JAGS-code in our R-program, we need a few more lines of code. The main choices there are for the number of MCMC-chains to run (we choose 3) and for the lengths of the MCMC before and after burn-in (we choose 10,000).

```
writeLines( ModelA, con="ModelA.txt" )
data.A <- list ( ny=ny, y=y, t=t )
ModelA <- jags.model( "ModelA.txt", data=data.A, n.chains=3, n.adapt=1e4 )
update( ModelA, n.iter=1e4 )
ModelA.coda <- coda.samples( ModelA, var=c("a","b"), n.iter=1e4 )
```

16.2.2 Model B: Cv-Specific Intercepts and Slopes, Not Hierarchical

The code for model B is almost the same except that the parameters a and b are now indexed such that the model estimates their values separately for the $ncv = 6$ different cultivars.

```
ModelB <- " model {
  for (i in 1:ny) {
    y[i]        ~ dnorm( a[cv[i]] + b[cv[i]] * t[i], tau.y[i] )
    tau.y[i] <- pow( sigma.y[i], -2 ) }
  for (c in 1:ncv) { a[c] ~ dnorm( 5e2, 1e-6 ) ; b[c] ~ dnorm( 1, 1e-2 ) } } "
```

16.2.3 Model C: Cv-Specific Intercepts and Slopes, Hierarchical

Finally we have the code for model C which begins the same as for model B, but adds lines for the hyperparameters. Note that we use normal distributions for the hyperpriors of μ_a and μ_b and uniform distributions for σ_a^2 and σ_b^2.

```
ModelC <- " model {
  for (i in 1:ny) {
    y[i]        ~ dnorm( a[cv[i]] + b[cv[i]] * t[i], tau.y[i] )
    tau.y[i] <- pow( sigma.y[i], -2 ) }
  for (c in 1:ncv) { a[c] ~ dnorm( mu.a, tau.a ) ; b[c] ~ dnorm( mu.b, tau.b ) }
  mu.a ~ dnorm(5e2,1e-6) ; tau.a <- pow(sigma.a,-2) ; sigma.a ~ dunif(0,1e3)
  mu.b ~ dnorm(1,1e-6)   ; tau.b <- pow(sigma.b,-2) ; sigma.b ~ dunif(0,2) } "
```

16.2.4   Comparing Models A, B and C

We now look at the results from all three models. Figure 16.3 summarizes the marginal
posterior distributions for the parameters and (for model C) hyperparameters.

First of all, we notice that the posterior uncertainty for the two parameters of model
A is very small, compared to the large variation in intercepts and slopes that we see for
models B and C. This is because all 258 data points provide independent information
for the estimation of just this single pair of parameters. We are reminded here that
very low uncertainty about the parameters of a model does not mean that we can be
confident that model structure is correct. Parameter uncertainty is generally smallest
in oversimplified models. If we now compare the differences between cultivars, as
determined by models B and C, we see much greater variation (especially for the
intercept parameter a) in the non-hierarchical model. This is because model B is
in fact a suite of six completely independent models, each of which only considers
its own subset of data points. In the hierarchical model C, we see smaller variation
between the cultivars: there is so-called *shrinkage* to the mean as defined by the
hyperparameters ξ_a and ξ_b. This is typical behaviour of hierarchical models and
it is what we want: hierarchical modelling has in-built protection against outliers.
The degree of shrinkage is highest for cultivars 1 and 6 because there were fewest
observations for them (10 and 14, whereas all others had more than 35). So the
BHM embodies the common sense that unusually high or low observations with
poor replication are likely over- or underestimates. The estimates for cultivars 1 and
6 in the BHM also show lower posterior uncertainty, they have *borrowed strength*
from the more frequently observed other cultivars.

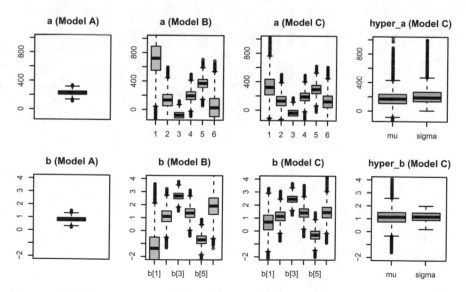

Fig. 16.3 Posterior parameter distribution for the three models

16.3   Applicability of BHM

BHM, as represented by model C, is an extremely useful approach to modelling systems where the individual components or members are not completely unrelated and dissimilar (model B), but are also not perfectly identical (model A)—and that actually covers almost any system you may be working on! Bayesian Hierarchical Modelling should be the default approach to statistical modelling. It frees the modeller from the need to define as constants any parameters that in reality are variable, without running the risk of overparameterisation. For further reading, the books by Gelman et al. (2013) and Gelman and Hill (2006) are recommended. When delving into the literature, do not be discouraged by the usual Tower of Babel: hierarchical modelling is the same as *multilevel modelling* and *mixed modelling*, and our main example model C is also referred to as a *varying-intercept, varying-slope model*.

Exercise

1. Taking the H out of BHM. In Chap. 8, we showed twelve different ways of carrying out a linear regression. Could we have added BHM to the list? What kind of prior distribution for hyperparameters would make a BHM essentially identical to a non-hierarchical model?

Chapter 17
Probabilistic Risk Analysis and Bayesian Decision Theory

As Bayesians, we try to acknowledge all our uncertainties about data and models, and express them as probability distributions. As we have seen in the preceding chapters, this approach allows us to quantify predictive uncertainty when using our models to predict the future. And this is of course important for the user of these predictions, whether that user is us or someone whom we report our results to. Our probabilistic results allow not just prediction but also calculation of risks and, more generally, support for decision-making. This chapter begins by explaining a rigorous method for analysing risks, and then broadens the perspective to Bayesian Decision Theory (BDT).

17.1 Risk, Hazard and Vulnerability

Risk is commonly defined as the expectation value for loss, irrespective of the system of interest. This means that we always distinguish two conditions: one where the system is not affected by some *hazard* and one where the system is being damaged and incurs a loss. Most risk theory is developed for discrete hazards that cause sudden system failure. Theoretical advances for systems where the hazard variable is always present and continuously varying, with matching continuous system response, are less well developed (Van Oijen and Zavala 2019). Dynamic systems performance varies with resource availability and other dynamic constraints, e.g. crop growth depending on water supply, or urban health as a function of air pollutant concentration. Risks from such continuous hazards are not associated with sudden discrete events, but with extended periods of time during which the hazard variable exceeds a threshold. To manage such risks, we need to know whether we should aim to reduce the *probability of hazard threshold exceedance*, or the *vulnerability of the system*.

There is only one possible definition of vulnerability that allows formal decomposition of risk (loss expectation) as the product of hazard probability and system vulnerability ($R = p[H] V$) (Van Oijen et al. 2013). The unique and unambiguous definition of vulnerability may resolve the general confusion in the risk literature

© Springer Nature Switzerland AG 2020
M. van Oijen, *Bayesian Compendium*,
https://doi.org/10.1007/978-3-030-55897-0_17

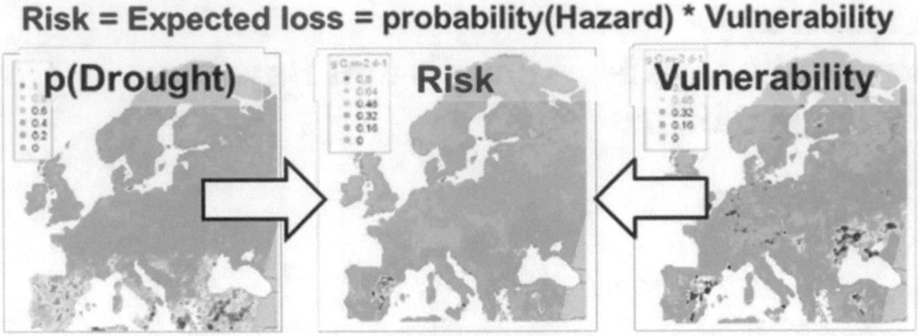

Risk = Expected loss = probability(Hazard) * Vulnerability

Fig. 17.1 Probabilistic risks analysis for drought impacts on European forest. Risks are highest in the Mediterranean area primarily because of greater hazard probability, but also because of greater vulnerability to drought. After Van Oijen et al. (2014)

about what 'vulnerability' means (Schneiderbauer and Ehrlich 2004). We used the decomposition-approach to analyse risks from summer droughts to the productivity of vegetation across Europe under current and future climatic conditions (Van Oijen et al. 2014). This analysis showed that climate change will lead to greatest drought risks in southern Europe, primarily because of increased hazard probability but also because of changes in vulnerability (Fig. 17.1).

17.1.1 Theory for Probabilistic Risk Analysis (PRA)

We now give a brief but formal outline of the PRA-method (Van Oijen et al. 2013, 2014). The method was designed to be widely applicable, allowing analysis of the response of any system variable z to any environmental variable x, univariate or multivariate. The domain of x is split in two regions, representing hazardous and non-hazardous conditions, typically separated by a threshold value of x. In the example given in the two papers, z stood for the productivity of vegetation, x was water availability (quantified as the Standardized Precipitation-Evapotranspiration Index or SPEI) and the hazard was drought, defined as SPEI being less than a given threshold. The probability of hazardous conditions (drought) was denoted as $p[H]$ and the probability of no drought as $p[\neg H] = 1 - p[H]$. We defined risk (R) as expected loss, i.e. the degree to which the occasional occurrence of hazardous conditions would make the average value of z less than potential: $R = E[z|\neg H] - E[z]$. Risk thus depends on both the probability of hazardous conditions and the damage then done. When both are high, so is the risk. The final term to be introduced was vulnerability V, which was implicitly defined by requiring that risk can be decomposed as: $R = p[H]V$, from which followed that $V = R/p[H] = E[z|\neg H] - E[z|H]$. No other definition of vulnerability would allow the risk to be written as the product of two distinct terms (one measuring the environment, the other the system response).

So vulnerability is average system performance (expectation of z) under "good" conditions (*not H*) minus that under "bad" conditions (H). Note that in this formalism, $p[H]$ is dimensionless, whereas R and V are intensive variables (defined per unit of area) that have the same dimension, e.g. $kg\,m^{-2}$.

The risk-decomposition has recently been expanded to include *exposure* (U), such that $R = p[H]UV$ (Van Oijen and Zavala 2019). The introduction of U allows upscaling of the risk analysis from individuals to exposed populations or from a point in space to regional exposure indices. Multiplication by U in units of area such as m^2 will convert R to an extensive variable (in units of say kg); V remaining intensive ($kg\,m^{-2}$).

Uncertainties arise from incomplete knowledge regarding hazards, exposures and vulnerabilities. Under specific conditions (e.g. temporal independence), uncertainty in hazard probability may be derived by treating the occurrence of hazardous conditions in the given scenario as drawn from a binomial distribution with proportion parameter $p[H]$. We can then quantify uncertainty in $p[H]$ using conjugate Bayesian updating starting from a beta-prior. In contrast, uncertainty in U and V is harder to quantify, as they are based on continuous rather than discrete probability distributions, and V is calculated as the difference between expectation values for overlapping (i.e. non-independent) distributions.

17.2 Bayesian Decision Theory (BDT)

The above PRA is useful to summarize the implications of our predictive uncertainty for risk, but for decision-making a wider framework is required, *Bayesian decision theory* (BDT) (Berger 1985; Jaynes 2003; Williams and Hooten 2016). BDT identifies the action portfolio that maximizes the *Bayesian expected utility* (or minimizes the *Bayesian expected loss*), where the utility *u(d,x)* is defined as the net benefit of action portfolio d under conditions x. The following very brief outline of BDT follows (Lindley 1991). Every decision problem has three main ingredients:

1. An *exhaustive* list of *exclusive* decisions $\{d_i\}$: exactly one from the list is taken.
2. An exhaustive list of events or environmental conditions $\{x_j\}$: more than one from the list can happen. The events are uncertain, so we have $\{p[x_j]\}$, possibly $\{p[x_j|d_i]\}$.
3. For every combination of d_i and x_j we assess the *utility* $u(d_i, x_j)$.

We can equally well develop the theory in terms of *loss* of utility (as is common in risk analysis), often denoted as $l(d_i, x_j)$, but choose utility here. The *expected utility* for a decision d_i is:

$$\bar{u}(d_i) = \sum_{j=1}^{n} p[x_j|d_i]u(d_i, x_j) \qquad (17.1)$$

The Bayesian solution to the decision problem is then simply: choose that decision d_i that has the maximum expected utility. But what is Bayesian about that? Nothing as such except that we require probabilities and utilities to be *coherent* so that we can use the rules of probability theory, including Bayes' Theorem, to manipulate them. This is especially important if we want to break down the decision problem in a sequence of smaller decisions with more easily quantified conditional probabilities, and when we are receiving new information that can help us update our probability distributions.

It is possible that there are infinitely many different possible choices, e.g. when the decision involves choosing the optimal value of a continuous variable. Then we do not have discrete but continuous probability distributions $p[x|d_i]$ and the above equation is written with an integral rather than a summation.

17.2.1   Value of Information

It would clearly be valuable to have reduced uncertainty about which of the uncertain events $\{x_j\}$ will occur, because then we could focus our utility calculation on those events. Let's consider the simple situation where only one of the $\{x_j\}$ (assumed independent of d_i) can occur but we do not know which one. How much would information be worth that tells us which specific event will happen? Lindley (1991) defined the *expected value of perfect information* ($V_{perfect}$) as the expectation, over $p[x_j]$, of the maximum possible utility under each event minus the maximum uninformed expected utility. In formula-form:

$$V_{perfect} = \sum_{j=1}^{n} p[x_j] \max_i u(d_i, x_j) - \max_i \sum_{j=1}^{n} p[x_j] u(d_i, x_j)$$

$$= \sum_{j=1}^{n} p[x_j] \max_i u(d_i, x_j) - \max_i \overline{u}(d_i) \tag{17.2}$$

Lindley also defined the *expected value of partial information*, that we denote as $V_{partial}$. In this case, the new information y_{new} does not identify the unique x_j but it does allow us to use Bayes' Theorem and modify the probabilities of the events from $\{p[x_j]\}$ to $\{p[x_j|y_{new}]\}$. To decide on how valuable that partial information is expected to be, we must sum over all possible values that the data $y_{new} \in Y_{new}$ can take (or integrate in the continuous case), which gives us:

$$V_{partial} = \sum_{Y_{new}} \max_i \sum_{j=1}^{n} p[x_j] \, p[y_{new}|x_j] \, u(d_i, x_j) - \max_i \overline{u}(d_i) \tag{17.3}$$

The above equations can help us decide whether we want to invest in acquiring extra information before we make our decision.

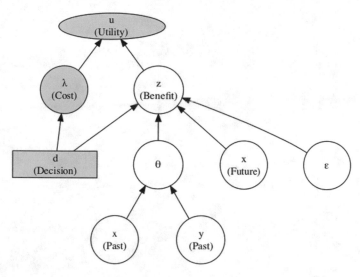

Fig. 17.2 A graphical model for Bayesian decision theory. See text for explanation of symbols

17.3   Graphical Modelling as a Tool to Support BDT

Let us depict the general decision problem graphically. We are working with uncertain quantities and conditional probabilities, so we can use the graphical modelling formalism of Chap. 15. That means that we design a *Directed Acyclic Graph (DAG)* that captures the decision-making joint probability distribution. The DAG is shown in Fig. 17.2.

Our graph helps clarify how we quantify $u(d, x)$. Neither actions d nor environmental conditions x directly affect u. Instead, the utility is the difference between the benefits from our system's future performance (z in the graph) and the costs λ. Both depend on our actions d and the system performance also depends on the environmental conditions (x_{Future}). Our conditional probability distribution for z will depend on model prediction $f(x, \theta)$ with associated error ϵ as we saw in the earlier chapters. Parameter uncertainty will be reduced after Bayesian calibration of the parameters on past observations (x, y). Lindley's 'value of information' discussed above pertains to reduction in our uncertainty about the environmental conditions x_{Future} that will partly determine system performance, besides our actions d.

In conclusion, risk analysis and decision theory bring together many of the themes of this book: modelling, data acquisition, uncertainty assessment, Bayesian calibration, graphical modelling. They are tools for making models practically useful.

Chapter 18
Approximations to Bayes

Bayes' Theorem tells us how to learn from data. We just need to assign our prior probability distribution for parameters or models and formulate the data likelihood function. The posterior distribution is then fully determined, and it encapsulates everything of interest. With the posterior in hand, we can make predictions with proper uncertainty quantification, we can carry out risk analysis and provide decision-support. So the basic ideas are extremely simple and powerful. But we have also seen that assigning a prior and likelihood is not always easy, and deriving a sample from the posterior distribution may require computationally demanding methods such as MCMC. So people keep searching for shortcuts where the Bayesian analysis can be made faster albeit perhaps a little bit less informative and accurate. Some of these approximation methods date from times when computers were slow and application of sampling-based Bayesian analysis was inevitably time-consuming. But the problem with computational demand has not gone away. The advent of complex computer simulators, such as global climate models (GCM) and dynamic vegetation models (DGVM), is keeping computational efficiency high on the agenda. There will always remain an important role in computational statistics for approximation methods. But I do believe that they are over-used. There is no better way of assessing your own understanding of a system than thinking carefully about all the system's parameters and about all measurements and their uncertainties. Writing down a prior and likelihood function is a learning experience in itself. And the posterior gives you more information than any of the outcomes from approximation can do. Another reason to prefer the real Bayesian deal over approximation methods is that Bayes' Theorem makes no assumptions about how many data you have (you don't need infinitely many), or about your parameter uncertainties (they need not be uniform or Gaussian) or your model (it need not be linear in any way). Bayes gives you the freedom to express exactly what you know and don't know. Bayes' Theorem can handle it.

This book focuses on explaining the basic applications of Bayesian analysis, so we shall not discuss any of the approximation methods at length. But here is very briefly a list of some of the most important ones with pointers to the literature.

© Springer Nature Switzerland AG 2020
M. van Oijen, *Bayesian Compendium*,
https://doi.org/10.1007/978-3-030-55897-0_18

18.1   Approximations to Bayesian Calibration

Methods for approximating and speeding up Bayesian calibration of model parameters include:

- *Approximate Bayesian Computation* (ABC). In ABC, you do not formulate a likelihood function. Instead you evaluate the likelihood of a parameter vector $p[y|\theta]$ by generating multiple different 'observations' from which you calculate summary statistics that you compare with the corresponding summary statistics of the real data to feed into an accept-reject rule. This is more suited to stochastic models or chaotic models with initial-state uncertainty than deterministic simulators. For more, see: Lintusaari et al. (2017).
- *Linear Bayes* (LB). In LB, the whole idea of formulating joint probability distributions is given up. The two central equations of LB, for adjusting expectation and variance, resemble the Kalman Filter equations. For more, see: Goldstein (2015). Nott et al. (2014) interpret ABC within the framework of LB.
- *Integrated Nested Laplace Approximation* (INLA). INLA makes assumptions of linearity and uses Laplace Approximations of probability distributions to arrive at marginal posterior distributions only, i.e. the distributions for individual parameters. No posterior parameter correlations are quantified. INLA has been adopted widely in the spatial modelling community. For more information, see: Blangiardo et al. (2013), Opitz (2017) and Bakka et al. (2018).

18.2   Approximations to Bayesian Model Comparison

Bayes' Theorem implies that the *integrated likelihood* (IL) is the main criterion for model comparison (see Chap. 12). However, the IL is difficult to quantify reliably, especially for high-dimensional parameter spaces. Various so-called *information criteria* have been proposed for model evaluation and selection, each of which can be seen as an approximation to the IL. One of the oldest and still most widely used is the *Akaike Information Criterion* (AIC) which is defined as follows:

$$AIC = -2\,log(\,p[y|\theta^{MLE}]) + 2\,n_p,$$

where θ^{MLE} is the maximum likelihood parameter estimate and n_p is the number of parameters in the model. Models with low AIC are preferred. But you notice the absence of a prior distribution in this definition and instead there is the penalty proportional to the number of parameters. The AIC only ranks models reliably when there are many data (so the prior is of little importance) and when the model is fairly linear—but those are conditions for which the IL can be easily calculated as well. Similar reservations can be expressed against the whole zoo of information criteria which nowadays include the WAIC, BIC, DIC and many more. For more information, see: Burnham and Anderson (2002), Bolker (2008), Hojsgaard et al. (2012) and Gelman et al. (2014).

Chapter 19
Linear Modelling: LM, GLM, GAM and Mixed Models

Science increasingly recognizes the nonlinearities in nature, and Bayesian methods can handle nonlinear models without any problem. However, linear modelling remains the default statistical approach for many, and it is therefore important to be familiar with the field. This chapter gives a brief overview of the most common classes of linear modelling.

19.1 Linear Models

Linear models are models whose output depends linearly on parameters. So the quadratic model $f(x, \theta) = \theta x^2$ is a linear model even though the dependence on x is not linear. More complicated models like $f(x, \theta) = \theta_1 x_1^2 + \theta_2 \, log(x_2 x_3)$ are also linear in their parameters. That shows that quite a wide variety of system behaviour can be represented by linear models. However, linear models are models of convenience. We do not use them because nature is linear, but because they are well-behaved and make data analysis easy. So they can be very useful in a first exploration of new data, but we should be careful when using them for prediction, especially when extrapolating the predictor variables x beyond values for which we have observations.

Before introducing different classes of linear modelling, we need to revisit our earlier discussion of the meaning of the term *model* (Chap. 2). Generally, in this book, we define a model as a function f of predictor variables x and parameters θ. But when we expand the model with a term that quantifies our uncertainty about measurement or modelling error, then we must distinguish between the *core model f* and the *statistical model $f + \epsilon$* (with $\epsilon \sim p[\epsilon]$). When people apply linear modelling in their work, they generally use the term 'model' to indicate the whole statistical model, and they refer to the core-model part as the 'linear predictor'. In short, a linear model is a statistical model that is written as the sum of a linear predictor f and an error term ϵ with an appropriate probability distribution assigned to the latter. For ease of discussion in this chapter, we shall consider the differences between a correctly

© Springer Nature Switzerland AG 2020
M. van Oijen, *Bayesian Compendium*,
https://doi.org/10.1007/978-3-030-55897-0_19

Fig. 19.1 Generalization of
the basic linear model (LM)
to generalized linear and
additive models (GLM,
GAM) and mixed models
(LMM, GLMM, GAMM)

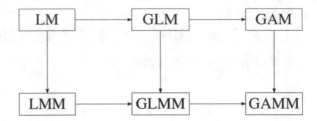

parameterised linear predictor and the observations y to be due to measurement error
(not model error), so we can write $y = f(x, \theta) + \epsilon_y$.

We shall restrict our discussion of linear modelling to the family tree depicted in
Fig. 19.1. The tree shows the six most common classes of linear statistical models,
from simple LM to more complex and flexible GLM and GAM—and their extensions to 'mixed models'. In all classes, the assumption is that the mean value of the
distribution $p[\epsilon]$ is zero, so the expectation value for as yet unobserved data Y is
$E[Y] = f(x, \theta)$.

19.2   LM

Linear statistical models (LM) are the first models that we consider here. In LM, the
linear predictor is a simple regression on the predictor variable(s), and the error term
has a zero-mean Gaussian distribution: $f(x, \theta) = X\beta$ and $\epsilon_y \sim N[0, \Sigma_y]$. We have
already seen examples of LM in this book, in the form of linear regression using
OLS, WLS and GLS (Chap. 8). These are examples of LM that only differ in the Σ_y
term.

19.3   GLM

Generalized Linear Models (GLM) differ from LM in both the linear predictor and
in the error term. In GLM, the linear predictor is wrapped in a transformation function and the error term need not be Gaussian but can be another distribution in the
exponential family of probability distributions which includes, besides the Gaussian,
distributions such as the Poisson, binomial, gamma etc. The transformation is generally applied to $E[Y]$ rather than to f, so the GLM is written as $g(E[Y]) = X\beta$,
where $g()$ is called the *link function*, with a choice for $p[\epsilon]$ that is appropriate for the
type of data y as well as the chosen link function g. A common example is *logistic
regression* which uses a logit link function and a binomial error distribution. Note
that if you choose the identity link function $(g(y) = y)$ and a Gaussian for ϵ, the
GLM reverts to a simple LM.

19.4 GAM

Generalized Additive Models (GAM) also use link functions but they define $g(E[Y])$ not as a linear regression $X\beta$ but as a sum of smooth functions $s(x)$, or a mixture of one or more regression terms plus one or more smooth functions. The 'smooths' can be functions of individual predictor variables or multiple variables, so statistical interactions can be represented. Typically, each $s(x)$ is written as the weighted sum of a number of *basis functions* which are often *splines*. So a GAM has many parameters that need to be estimated. But the fact that a GAM is still a linear model with respect to the basis function weights facilitates model fitting.

19.5 Mixed Models

Mixed models are hierarchical models. This means that at least for some of their parameters we not only estimate the values of the parameters themselves, but also properties ('hyperparameters') of the population that they belong to. A simple example would be a varying-intercept, common-slope model $y_i = \theta_i + x_i\beta + \epsilon$, with $\theta_i \sim N[0, \sigma^2]$. This is a linear mixed model (LMM) in which the hyperparameter σ^2 represents the variance of the intercept in the whole population. In typical confusing linear modelling jargon, the θ_i would be referred to as 'random effects' and the β as a 'fixed effect', but of course there is nothing random about the former and nothing fixed about the latter. There are just different degrees of, and representations of, our uncertainty. In practical use of mixed models, random effects are often nuisance parameters that people are not interested in (so a summary in terms of their variance is enough) whereas the fixed effects quantify the impact of predictor variables that are of primary interest to the researcher. You may for example be interested in the impact of a treatment variable like resource availability, but not in how differences between experimental blocks, individuals or locations affect that impact. By treating the latter differences hierarchically you quantify their variability. You can then account for that variability as a known source of uncertainty when making predictions of the treatment effect on new members of the same population.

Similar extensions to hierarchical modelling can of course also be applied to GLM and GAM, which then become GLMM and GAMM (e.g. Diggle and Ribeiro 2007; Gelman et al. 2013; Wood 2006).

19.6 Parameter Estimation

The main motivation for linear modelling is its simplicity and concomitant ease of parameter estimation. Most studies that use linear modelling estimate parameter values by means of (restricted) maximum likelihood estimation rather than full

Bayesian calibration. For the simple LM, there are analytical solutions (Lindley and Smith 1972), as we discuss in Chap. 8, but when we move to the more flexible models in our family tree, the methods for parameter inference become increasingly approximate, especially with small data sets (Wood 2006). The flexible models also run the risk of following the data too closely, making the response surfaces very wiggly. To avoid such tracking of noise rather than signal, we may want to introduce a 'penalty' to high curviness of the function that we want to fit, e.g. by reducing the likelihood value by the average second derivative of the function. In such *penalized regression*, maximum likelihood estimation is replaced by identifying the maximum of (something akin to) likelihood minus penalty. From a Bayesian perspective, the penalty function constitutes a prior on function shape. A more principled approach is to go full Bayesian and ensure that prior and likelihood are expressed as proper probability distributions. But if we do so, the beautiful theory developed for linear modelling will not help us much anymore: we will quickly need to resort to MCMC. For the mixed models, which are hierarchical, Gibbs sampling, or other MCMC algorithms that include a Gibbs-step, will be appropriate.

19.6.1   Software

Models in the class of LM can be fit analytically, but maximum-likelihood parameter estimation of LM and GLM can also be done with the basic R-functions `lm` and `glm`. For LMM and GLMM there is the R-package `lme4` , and for GAM and GAMM you can use the package `mgcv`, with `gamm4` as a more recent alternative. However, you can obviously always write your own MCMC code to do a full Bayesian analysis, which means adding a proper prior for parameters rather than using the uniform distribution that is implicitly chosen when you carry out maximum likelihood estimation. You will then also have the greatest freedom in choosing your own functions, including smooths, and probability distributions.

Chapter 20
Machine Learning

Machine learning is the name for a very wide collection of techniques for exploring data and estimating functions. The field has expanded and diverged to the extent that it is hard to find common denominators. But we can say that the majority of machine learning techniques do not focus on explanation or on uncertainty quantification, but on prediction. Machine learning is typically applied in cases where we have many data but not much understanding of causal pathways to guide our modelling.

The field has developed with more input from computer scientists than from statisticians or natural scientists. This may be surprising because machine learning is the offshoot from artificial intelligence research that aimed to build process-based models that mimic the behaviour of the human brain, with *neural network models* as the classical example. Despite this early aim of understanding how humans learn, many researchers use machine learning as a black box technique whereby the machine—rather than the human —learns to develop predictive algorithms from data. This has given us a problem with interpretability. You can open the black box model, but when you peek inside, you won't find anything resembling real-world processes. The algorithms developed by means of machine learning can be so complicated that we now see researchers in their turn developing computer algorithms for interpreting the computer-generated algorithms! An important motivation for developing these interpretation techniques is that machine learning algorithms are increasingly used in human society and need to be challenged (O'Neil 2016). The machine learning interpretation algorithms that have been developed so far are predominantly different forms of sensitivity analysis that help us identify the main predictor variables and interactions. It remains hard to derive ideas about causal pathways from these algorithms.

Besides the problem of poor interpretability, the second main problem of machine learning is that the methods are not geared toward uncertainty quantification. Machine learning uses training data to maximize the predictive capacity of highly flexible functions, trying to find unique optimal parameter vectors rather than posterior probability distributions. It has long been recognized that there is much scope for Bayesian methods to improve on this way of working (MacKay 1995), but progress has been limited.

© Springer Nature Switzerland AG 2020
M. van Oijen, *Bayesian Compendium*,
https://doi.org/10.1007/978-3-030-55897-0_20

20.1   The Family Tree of Machine Learning Approaches

Machine learning methods are generally classified according to the family tree of Fig.
20.1. The first division is between *supervised* and *unsupervised* learning (Murphy
2012). The terms are perhaps misleading—human intervention is always necessary—
but they refer to the nature of the data that the methods operate on. In the case
of supervised learning, the training data are (x, y), i.e. both predictor and dependent variables, whereas unsupervised methods only explore independent variables
x. Supervised learning derives algorithms for predicting y whereas unsupervised
learning identifies patterns in x. The supervised methods are subdivided further into
methods for *regression* and *classification*. The only difference is that in regression
the dependent variable y is continuous whereas in classification it is discrete and can
be associated with class labels. Various machine learning methods can be used for
both purposes. Unsupervised methods are subdivided into methods for *clustering* the
data x into different groups, and methods for *dimensionality reduction* which find
low-dimension projections of x that preserve most of the information in the data.

We can view the four classes of machine learning as attempts to find optimal functions $f(x, \theta)$ for different purposes. The classes differ in the nature of the training
data and in the constraints that they put on the shape of $f()$. In supervised learning, we use highly flexible functions $f()$ such that they can reproduce any kind of
pattern existing in the training data (x, y). Typical examples are neural networks
for regression, and *decision trees* (Chipman et al. 2010), *random forests* (ensembles
of decision trees) and *support vector machines* for classification—but the regression
methods can be used for classification and vice-versa (James et al. 2013). We shall discuss neural networks in more detail below to illustrate some of the key strengths and
weaknesses of machine learning. In unsupervised clustering, we try to find a function
$f(x, \theta)$ that assigns labels to the x such that distances between cluster members are
smaller than distances between clusters, with many different options for the choice

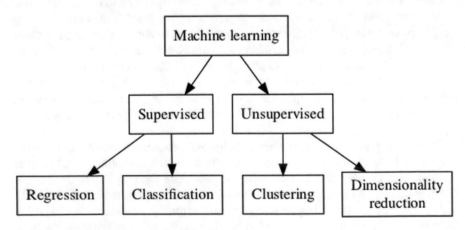

Fig. 20.1   Family tree of machine learning methods

of distance function. Finally, the main example of dimensionality reduction is principal component analysis (PCA) where we sequentially identify orthogonal linear data transformations ('components') $f_i(x, \theta) = x\theta$ such that $\Sigma\theta = 1$ and $Var[f_i]$ is maximized. By keeping only the first few components, which together account for most of the variance in the data, we reduce the dimensionality of x.

In practice, the role of the unsupervised learning techniques is mainly in the initial phase of studies, where we want to reduce the complexity of our data. The clusters and the transformed data of lower dimension can then be passed on to the supervised methods.

20.2   Neural Networks

Neural networks are functions. They take uni- or multivariate x as input and produce uni- or multivariate output. They are generalizations of regression models such that a function of any shape can be approximated to any desired precision with a neural network. An example of a simple neural network is shown in Fig. 20.2.

In the neural network of the figure, there is a layer of network nodes between the inputs and outputs—we say that there is one 'hidden layer' of neurons. In *deep learning*, more layers will be present. Our figure shows a 2-3-2 network, with two inputs, three neurons and two outputs, but everything upward from 1-1-1 is possible. So what function does our network represent, and what happens at the neurons? Well, each neuron is a small model in its own right that (1) calculates a weighted sum of its inputs, (2) adds a 'bias', (3) transforms the result. In other words, each neuron is

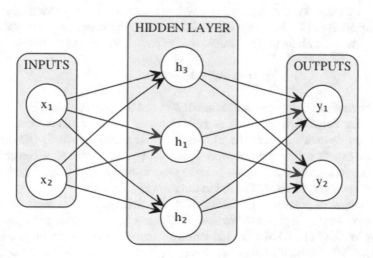

Fig. 20.2   A neural network with one hidden layer. The red arrows emphasize that each neuron is connected to all inputs and to all outputs and there are no loops, making this a standard feedforward neural network

a generalized linear model (GLM; see the preceding chapter on linear modelling). The transformation function is, in analogy to neuron activity in the brain, called an *activation function*. Activation functions are generally chosen to be nonlinear and often S-shaped such as the logistic or the hyperbolic tangent (written as `tanh()` in R). If you choose the former, your neurons will be mimicking logistic regression. If we assume that the latter function was used in our network, then the mini-model represented by the neuron labelled as h_1 is:

$$
\begin{aligned}
h_1 &= tanh(w_{11}x_1 + w_{12}x_2 + bias_1) \\
    &= tanh(\Sigma w_{1i}x_i + bias_1) \\
    &= tanh(X\theta_1),
\end{aligned} \tag{20.1}
$$

where $X = (1, x_1, x_2)$ and $\theta_1 = (bias_1, w_{11}, w_{12})^\top$. Let's now turn to the output layer. The output nodes are the same kind of mini-models as the hidden-layer neurons, but they have their own weights, biases and activation function. Taking everything together we can write the function expressed by neural networks like this one very compactly as follows:

FEEDFORWARD NEURAL NETWORKS:

$$
y_k = f^{(2)}(H\theta_k^{(2)}), \text{ where} \tag{20.2}
$$
$$
H = (1, f^{(1)}(X\theta_1^{(1)}), ..., f^{(1)}(X\theta_{n_H}^{(1)}),
$$

where n_H is the number of nodes in the hidden layer, and the $^{(1)}$ and $^{(2)}$ refer to neurons and outputs, respectively. The main point of writing the network function out like this is to make clear that a very complex-looking neural network is still just a function with inputs and parameters, so we can treat it like every other model in this book. Equation (20.2) shows the network function of so-called *feedforward neural networks* with one hidden layer. The number of parameters in such networks is equal to

$$
n_p = n_h(n_x + 1) + n_y(n_h + 1),
$$

where the first term gives the dimension of $\theta^{(1)}$ and the second that of $\theta^{(2)}$.

For other network types, such as *convolutional networks* (where each neuron depends on its own subset of the x) and *recurrent neural networks* (which allow loops), we cannot write out the network function in such a simple way. Nevertheless, even those networks are functions, just like process-based models are functions that we represent algorithmically and not in the form of a formula.

Let us now examine the flexibility of neural networks: what range of system behaviours can they represent? We shall implement in R a similar example as was given by MacKay (1995) of a one-hidden-layer neural network with univariate input and output. We choose the hyperbolic tangent ($f^{(1)}(x) = tanh(x)$) as the activation function for the neurons, and the identity function ($f^{(2)}(h) = h$) for the outputs. The hyperbolic tangent function is defined as $tanh(x) = \frac{e^x - e^{-x}}{e^x + e^{-x}}$ so it maps input

$x \in [-\infty, \infty]$ to $[-1, 1]$. It is a scaled version of the standard logistic $g(x) = \frac{1}{1+e^{-x}}$ (which maps to $[0, 1]$) such that $tanh(x) = 2g(2x) - 1$. The size of our one-layer network is determined by the number of neurons in the single hidden layer. The network function depends on the weights and biases that we set for each neuron and for the single output. We could implement this specific network in R with very short code: `y <- sum( tanh(x*w1+b1) * w2 ) + b2`, but we'll make the code more general such that it can produce any single-hidden-layer univariate I-O feedforward neural network, and can also accept a vector for x so that multiple input values are processed in parallel.

```
fNN.11 <- function( x, w1, b1, w2, b2, f1=tanh, f2=identity ) {
  X <- cbind(1,x) ; theta1 <- rbind(b1,w1) ; theta2 <- c(b2,w2)
  H <- cbind(1, f1(X
  y <- f2(H
  return( y ) }
```

You see in the R-code that we choose the hyperbolic tangent as the default activation function for the neurons (`f1`) and the identity function for the output node (`f2`), but we give no default values for inputs `x`, weights `w1`, `w2` or biases `b1`, `b2`. The number of neurons is implicitly defined in this code as the length of the vectors `w1`, `b1` and `w2`, so you can have any number of neurons in the hidden layer that you wish.

This type of neural network, with just one hidden layer, is very flexible and can represent any kind of continuous function, as long as we have enough neurons. Figure 20.3 shows examples of network functions generated by a 1-400-1 network. We assigned a zero-mean Gaussian prior to all parameters (weights and biases) with variances for neurons equal to 100 and for outputs equal to 1. From this prior, we generated a hundred different network functions—the figure shows three of them as well as the sample mean and standard deviation. Note how strongly the three network functions differ and their complicated shape.

20.2.1 Bayesian Calibration of a Neural Network

Let's now carry out Bayesian calibration of a neural network on some simple test data $x = (-1, 0, 1)$, $y = (6.09, 8.81, 10.66)$. We only use 5 neurons this time to keep the number of parameters to calibrate manageable. We use the Metropolis algorithm for MCMC, with the same R-implementation that we introduced before. The MCMC-generated sample from the posterior distribution for all network parameters is shown in Fig. 20.4.

The top row of Fig. 20.5 shows a few network functions of this 5-node network. The functions were sampled from the prior and posterior and also shown are the network functions associated with the prior and posterior mean. Note that the mean parameter vector does not fit well, as we would expect because the neural network is nonlinear. More representative than the mean is the parameter vector with highest posterior probability (the 'MAP' vector) whose predictions are shown too.

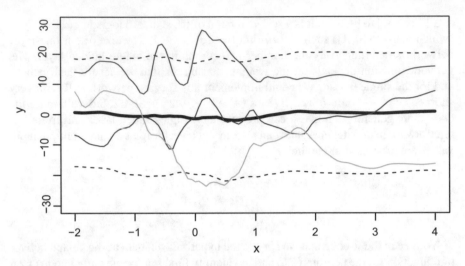

Fig. 20.3 Sampling from a neural network with 400 neurons. Black lines: mean of 100 network functions sampled from the prior +− standard deviation. Non-black lines: three of the sampled network functions

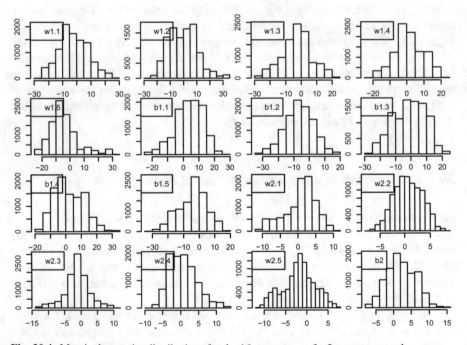

Fig. 20.4 Marginal posterior distributions for the 16 parameters of a 5-neuron network

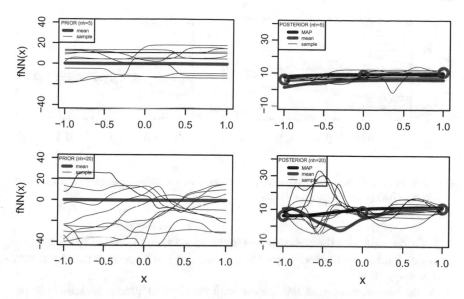

Fig. 20.5 Prior and posterior predictions of different one-hidden-layer neural networks. Top row: 5 neurons. Bottom row: 20 neurons

We repeated this exercise with a 20-node network, and the results are shown in the bottom row of the same Figure 20.5. We see that the network with 20 nodes expresses greater prior uncertainty about the shape of the network function than the 5-node network. With 20 nodes, we already find considerable wiggliness, which may be undesirable if we believe that the true relationship between x and y is not so wiggly. And much of the prior wiggliness carries over to the posterior: when we look at the sampled functions from the posterior distribution for the network parameters, we see again that the network functions are less smooth with more neurons. The use of neural networks, and much of machine learning in general, is a balancing act between desired flexibility and unwanted overfitting.

20.2.2 Preventing Overfitting

There are four obvious ways to constrain the behaviour of neural networks: (1) get more data, (2) restrict the values that the parameters can assume, (3) add a penalty term to the likelihood function that makes wiggly network functions unlikely, (4) limit the number of nodes and layers. All four ways have been tried with neural networks, but usually by means of ad-hoc methods (under headings such as 'regularization' and 'penalization') rather than Bayesian probability theory (James et al. 2013; MacKay 2003).

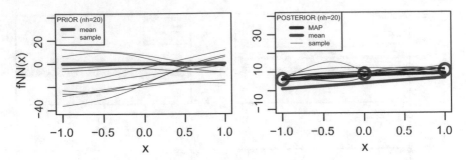

Fig. 20.6 Prior and posterior predictions of a 20-neuron one-hidden-layer neural network with narrow prior for neuron weights and biases

Let's examine what the consequences in the above example would be if we set the prior distribution for the neuron parameters (w1, b1) to be zero-centered and very narrow (Fig. 20.6).

As the figure shows, with this narrow prior the network produces mainly smooth functions, both a priori and a posteriori, without sacrificing its capability to reproduce the training data. But remember that our data set is atypically small and well-behaved. With richer data sets, we will need a richer network with more neurons. However, a full Bayesian approach to derive (a representative sample from) the posterior distribution will then become difficult. Neural networkers therefore usually go no further than estimating the single parameter vector that is optimal for their training data. They use optimisation algorithms specifically designed for the different types of network, such as the *backpropagation* gradient-descent algorithm for feedforward networks with multiple hidden layers.

20.3 Outlook for Machine Learning

Our simple example of neural networks showed how machine learning starts from highly flexible function definitions which it aims to constrain using data. We could have chosen other machine learning formalisms, such as random forest modelling, to illustrate the flexibility as well as the associated problems of parameterisation. In machine learning, these problems have often been swept under the carpet by not attempting to derive a full posterior probability distribution for the parameters, but instead just identifying a single well-performing parameter vector through minimization of a *cost function* (pseudo-likelihood). Parameter uncertainty is thus not assessed, but predictive uncertainty can still be approximately quantified using cross-validation (splitting the data into training and validation data, in various configurations). Despite the continuing prevalence of these ad-hoc methods for parameterising machine learning models, Bayesian methods are slowly being introduced. There are, for example, R-packages for Bayesian regression tree modelling (Kapelner and Bleich 2016) and in 'caret', a powerful general-purpose R-package for machine learning, various Bayesian methods are included (Kuhn 2008).

Are there not other functions that are as flexible as the typical machine learning functions yet facilitate uncertainty quantification? Well, at various places in this book (e.g. Chap. 14) we have encountered *Gaussian Processes* (GP) which seem to fit the bill. In fact, Neal (1996) showed that the prior predictive uncertainty represented by the output from an untrained feedforward neural network with Gaussian priors for the weights and biases (as was used to produce Fig. 20.3) will—in the limit of an infinite number of neurons—be equal to the predictive uncertainty of a GP. However, calculations with GP require the computationally demanding procedure of matrix inversion, so they may not always be a good substitute. Bayesian machine learning remains underdeveloped.

At the beginning of this chapter, we mentioned another problem with machine learning: it is a black-box approach that leads to algorithms that are difficult to interpret. To ameliorate the problem there are efforts to insert mechanistic modelling into some of the layers of deep learning networks (Reichstein et al. 2019). Essentially, this is hybridizing machine learning with process-based modelling. We generally use process-based modelling where we understand a system's mechanisms well, whereas machine learning is used where we have many data but little understanding. Hybridizing the approaches is a highly intriguing idea that will hopefully lead to algorithms that are not only easier for humans to understand but also have greater predictive capacity in conditions that differ from those prevailing during collection of training data.

We can also imagine going beyond hybridization, and replacing the process-based models completely by machine-learned algorithms: model emulation. We briefly discuss data-driven versus mechanistic model emulation in Chap. 14. A more modest goal would be to only represent model structural error ('discrepancy') by means of a machine-learned algorithm (see Chap. 13).

Exercises

1. The neural network of Fig. 20.2. How many parameters does this neural network have? Use the formula provided in this chapter.

2. 1-1-1 neural networks. The simplest neural networks are 1-1-1 networks. How could you specify in R a 1-1-1 network that just doubles its input? Use the R-function fNN.11 that we defined in the chapter.

Chapter 21
Time Series and Data Assimilation

Time series modelling can be defined as modelling the progression over time of the state of a dynamical system. We focus here on simple statistical models rather than complex process-based models (PBMs), but will come back to those at the end of this chapter (and for more about PBMs see e.g. Chaps. 9, 12 and 13).

From a Bayesian perspective, we always express what we want to predict as a conditional probability distribution. The future is conditional on the present, and our uncertainties about the present will affect the quality of our prediction. So all time series modelling is a matter of specifying the probability distribution $p[future|present]$ or $p[z(t + 1)|z(t)]$. And in common time series practice, uncertainties tend to be expressed as Gaussian distributions. That leads to the question: can we use Gaussian Processes, which we encountered a few times already, to represent common time series models? Let's begin there.

21.1 Sampling from a Gaussian Process (GP)

We generally use a GP to represent our uncertainty about the shape of a continuous function f (Chap. 14). The domain A of f can be unbounded and multidimensional ($A = \mathbb{R}^n$) or just a line segment ($A \subset \mathbb{R}$). In all cases A is an infinite set of points over which the GP defines a joint Gaussian probability distribution for $f(s), s \in A$. The GP is fully defined by its mean and covariance functions $m(s)$ and $C(s, s')$. But how exactly does the GP represent our uncertainty about $f(s)$? How can we produce a representative sample of possible function shapes? We can do this step for step by first sampling the GP at a single point, then at another conditional on the value we sampled for the first point, and so on. That iterative sampling procedure uses the decomposition that we can write for every joint distribution: $p[F] = p[f_0] \, p[f_1|f_0] \dots p[f_n|f_0, \dots, f_{n-1}]$. We can of course only sample at a finite number of points, so this does not fully determine the shape of the function.

© Springer Nature Switzerland AG 2020
M. van Oijen, *Bayesian Compendium*,
https://doi.org/10.1007/978-3-030-55897-0_21

How many points we need for a good approximation depends on the expected wiggliness of the function which is determined by the covariance function $C(s, s')$. Let's give an example for a function f whose one-dimensional domain A is the interval from 0 to 30. We represent our uncertainty about f as a GP with mean function $m(s) = 0$ and covariance function $C(s, s') = 3 \, exp(-|s - s'|/10)$. This choice of GP means that we expect $f(s)$ to be centered around zero with an overall variance of 3 and correlations between points that decrease exponentially with distance at a correlation length of 10. This is a much simpler GP than we introduced in the chapter on model emulation but we could still use the general R-function for GP-prediction that we implemented in that chapter and which we named `GP.pred`. But let's write, for present use, a simplified version of that R-function so you do not need to look it up (although you might find it a good exercise to compare the two versions):

```
GP.pred.0 <- function(x0,x,y,Sy,phi) {
    C0    <- Sy[1] * exp( -abs(x-x0)/phi )
    m0_y <- t(C0) %*% solve(Sy) %*% y
    S0_y <- Sy[1] - t(C0) %*% solve(Sy) %*% C0
    return( list( "m0_y"=m0_y, "S0_y"=S0_y ) ) }
```

Now we are ready to start sampling from our GP. We assume that the domain of the function, which is the continuous interval [0,30], will be adequately covered if we sample at $x = 0, 1, ..., 30$. We start at $x_0 = 0$, where we generate a value for $y_0 = f(x_0)$ from the Gaussian $N[m(0), C(0, 0)] = N[0, 3]$ and then we keep applying our R-function `GP.pred.0` to sample each following point. When we have reached x_{30} and produced a value for y_{30} we will have produced one sample 'function'. To sample representatively from our GP we do this 1000 times. The following R-code does the sampling.

```
set.seed(13)
Vy   <- 3 ; phi <- 1e1 ; nseries <- 1e3 ; n <- 30
Y           <- NULL
for(i in 1:nseries) {
    x    <- 0 ; y  <- rnorm(1,0,sqrt(Vy))
    for(xi in 1:n) {
        dx <- as.matrix( dist(x) ) ; Sy <- exp( -dx/phi ) * Vy
        py <- GP.pred.0(xi,x,y,Sy,phi)
        yi <- rnorm(1, mean=py$m0_y, sd=sqrt(py$S0_y))
        x  <- c(x,xi) ; y <- c(y,yi)
    } ; Y <- cbind( Y, y )
} ; mY <- rowMeans(Y) ; sY <- apply(Y, 1, sd)
```

The results of the GP-sampling are summarized in the left panel of Fig. 21.1. As expected, our 1000 functions are mostly close to zero, with a standard deviation of $\sqrt{3}$. The individual functions in our sample, of which three are shown, differ a lot from each other, but all are quite wiggly.

You may well ask, what has all that to do with time series? Well, if we reflect on how we generated each sampled function, our procedure mimicked a temporal process. We kept moving our function evaluation from one point on the x-axis to the next. And that is what statistical time series are: they are rules for generating the value of a function $z(t)$ as a function of the value the function had in the past, e.g. $z(t - 1)$. Specifically, the time series mimicked by our GP is called an *autoregressive* time series with dependency on the last value, denoted as an *AR(1)* process. The general formula for an AR(1) process is:

AR(1) time series model:

$$z(t) = \alpha\, z(t-1) + \epsilon(t) \quad \text{where} \quad \epsilon(t) \sim N[0, \sigma_\epsilon^2].$$

(21.1)

If $|\alpha| < 1$, the AR(1) model is stationary such that our uncertainty about any future value of z is independent of t and captured by a Gaussian distribution: $z(t) \sim N[0, \sigma_\epsilon^2/(1-\alpha^2)]$. And the AR(1) model then has its own covariance function $C_{AR(1)}(t, t-k) = \alpha^k \sigma_\epsilon^2/(1-\alpha^2)$ (from which of course the variance follows by plugging in $k = 0$).

We now see that the AR(1) is equal to our previous GP if both use the same covariance function. We achieve that by setting $\alpha = exp(-1/\phi)$ and $\sigma_\epsilon^2 = 3(1-\alpha^2)$. We can then replace the GP-code with the simpler AR(1)-code without changing the underlying statistical model. Here is the code for sampling 1000 time series from the AR(1) process that should be equivalent to our GP:

```
set.seed(13)
alpha <- exp(-1/phi) ; Ve <- Vy * (1-alpha^2)
Z <- NULL
for(i in 1:nseries) {
  z <- rnorm( 1, 0, sqrt(Ve/(1-alpha^2)) )
  for(t in 1:n){
    z1 <- rnorm(1, mean=alpha*tail(z,1), sd=sqrt(Ve))
    z  <- c(z,z1)
  } ; Z <- cbind( Z, z )
} ; mZ <- rowMeans(Z) ; sZ <- apply(Z, 1, sd)
```

This is much faster code because it does not involve inversion of an ever-expanding covariance matrix for already examined points, as we have in the GP sampling. A summary of the 1000 time series produced by our AR(1) model is shown in the right panel of Fig. 21.1. The results are in every detail identical to the GP-results on the left! This is because the underlying mathematics is equivalent and the sampling for both models, which involved 31,000 calls to R's random-number generator (as you can verify from the code), was initialised with the same seed.

The AR(1) model can easily be made more general. We can add a non-zero intercept c and extend the memory of the system further back than the preceding step. That gives us AR(p) models, defined as $z(t) = c + \sum_{i=1}^{p} \alpha_i z(t-p) + \epsilon(t)$. We can also introduce a memory of the noise term, which leads to *moving average* time series models, denoted as MA(q). Combining the two model types gives us the quite flexible ARMA(p,q) models, defined as:

ARMA(p,q) time series model:

$$z(t) = c + \sum_{i=1}^{p} \alpha\, z(t-p) + \sum_{j=1}^{q} \beta_j\, \epsilon(t-q) + \epsilon(t) \quad \text{where} \quad \epsilon(t) \sim N[0, \sigma_\epsilon^2].$$

(21.2)

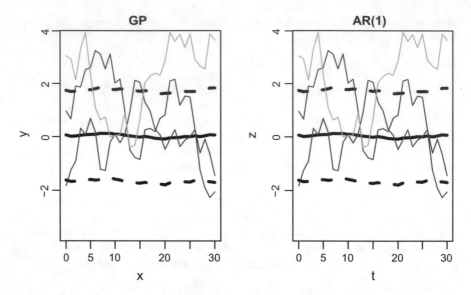

Fig. 21.1 Left: sampling from a Gaussian Process. Black lines are mean plus-minus one standard deviation of 1000 sampled functions for each value of x. Coloured lines show three sampled functions. Right: as on the left but from an AR(1) process rather than a GP. The same random seed was used for both panels

All these can be seen as examples of *state-space models* that are special cases of GP (Williams and Rasmussen 1996). This is just a reminder of the fact that a huge part of statistical modelling consists of combining linear relationships with Gaussian probability distributions in different ways. And it means that time series modelling does not add new ideas beyond what was already implicit in the GP that we have found useful in various applications. It all boils down to different ways of specifying and using the covariance function for a multivariate Gaussian distribution. However, time series modelling can be a very efficient choice, as we saw in the much higher speed of the AR(1) code compared to its equivalent GP counterpart. And the efficiency of time series modelling is maintained when we extend it with assimilating information from observations at each time step, as we investigate in the following section. That will also bring us back to truly Bayesian methods.

21.2   Data Assimilation Using the Kalman Filter (KF)

Data assimilation is the name commonly given to methods that keep alternating between predicting the future state of a system and correcting the state estimate using new data. So data assimilation uses data to learn about the *state* of a system, whereas Bayesian calibration uses data to learn about the *parameters* of a model.

The distinction is not very important as the underlying mathematics is the same, both involve modelling, and data assimilation algorithms tend to be Bayesian too, as we shall see.

Let's introduce some notation. In data assimilation, the future state z of a system is predicted using a model f, followed by using data y to correct that estimate. If the f in question is a state-space model, then data assimilation is a natural extension to the time series models that we have discussed in this chapter. Another term for data assimimilation of time series models is *filtering*. And if we stay in the Gaussian world of linear state-space modelling then the classical example of data assimilation is *Kalman filtering* (KF).

The KF in its richest formulation is a versatile data assimilation tool, but we begin by presenting a simplified KF that can only be used when the system z is univariate (i.e. at any time t, the system state $z(t)$ is a scalar, not a vector). We also assume for now that the state is directly observable, albeit noisily: $y(t) \sim N[z(t), \sigma_y^2]$. Finally, we assume that each application of model f multiplies the last state estimate by a constant α and adds Gaussian uncertainty with constant variance σ_f^2. The KF is then defined as follows:

KALMAN FILTERING (simple case of observable scalar state):

$$\text{'Kalman Gain': } K = \sigma_p^2 / (\sigma_p^2 + \sigma_y^2).$$

Data assimilation step: $p[z_a | \mu_p, \sigma_p^2, y, \sigma_y^2] = N[\mu_a, \sigma_a^2]$,

$$\text{where } \mu_a = \mu_p + K(y - z_p) \text{ and } \sigma_a^2 = (1 - K)\sigma_p^2.$$

Prediction step: $p[z_p | \mu_a, \sigma_a^2] = N[\mu_p, \sigma_p^2]$,

$$\text{where } \mu_p = \alpha \mu_a \text{ and } \sigma_p^2 = \alpha^2 \sigma_a^2 + \sigma_f^2.$$

(21.3)

In Eq. (21.3), I have left out the time indices to avoid cluttering, but don't forget to read the equations as steps that we keep repeating as we move through time. The following R-code implements this simple-case KF-algorithm, with α set to 0.9, the model variance σ_f^2 set to 1, and some arbitrary data added.

```
set.seed(13)
n        <- 30 ; y <- runif(n,0.8,1) ; Vy <- 3
zp       <- Vp <- za <- Va <- rep(NA,n)
zp[1] <- 0 ; Vp[1] <- Vf <- 1 ; alpha <- 0.9
for(i in 1:n) {
  Ki       <- Vp[i] / ( Vp[i] + Vy )
  za[i]    <- zp[i] + Ki * ( y[i] - zp[i] )
  Va[i]    <- ( 1 - Ki ) * Vp[i] ; if(i==n) break
  zp[i+1] <- alpha    * za[i]
  Vp[i+1] <- alpha^2 * Va[i] + Vf }
```

The results of this simple KF are shown in Fig. 21.2. The figure demonstrates some quite general properties of data assimilation. We see that after each model prediction of the system state ($z_p \sim N[\mu_p, \sigma_p^2]$), the assimilation of a new data point acts as a correction on that estimate, so that the $\mu_a = E[z_a]$ lie between model estimate μ_p and observation. We also see that our uncertainty about the state estimate is lowest after a data assimilation ($\forall t : \sigma_a^2(t) < \sigma_p^2(t)$). That is because the predictions

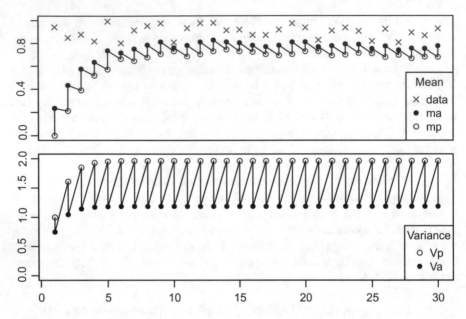

Fig. 21.2 Data assimilation using the KF. The top panel shows the observations and the alternating state estimates from model prediction (mp) and data assimilation (ma). The bottom panel shows the corresponding uncertainties as variances. Observational uncertainty is not shown but has a constant variance equal to 3

add model uncertainty. If observational and model uncertainties are constant as here (σ_y^2 and σ_f^2 are independent of t and z), then the variance of our state estimate will gradually start oscillating between constant values for σ_a^2 and σ_p^2. This happens because the repeated increase from σ_a^2 to σ_p^2 ($= (\alpha^2 - 1)\sigma_a^2[i] + \sigma_f^2$) becomes equal to the decrease from σ_p^2 to σ_a^2 ($= K_i \, \sigma_p^2[i]$). The fact that variable data keep coming in does not stop the variances from stabilising because—as you can see from Eq. (21.3)—the (changes in) variances do not depend on the *values* of the observations, only on their *uncertainty* which is the constant σ_f^2.

So how would the KF behave if our data were not very reliable? Well, we can check what happens when the data are not informative without changing our R-code by setting the data variance σ_y^2 to a very large value, say $\sigma_y^2 = 10^8$. If we do that, and also set $\alpha = exp(-1/10)$ and $\sigma_f^2 = 3(1 - \alpha^2)$, then we are back to our GP & AR(1) example of earlier in this chapter, and the variances then become the same as we found there (Fig. 21.1): $\forall t : \sigma_p^2(t) = \sigma_a^2(t) = 3$. This confirms that KF is just time series modelling extended with data assimilation at each time step.

21.2.1 A More General Formulation of KF

We now look at a more complete formulation of the KF: see Eq. (21.4). The first difference with our earlier, simplified implementation of the KF (Eq. (21.3)) is that all quantities are now vectors and matrices rather than scalars. So the state of the system $z(t)$ and the observations $y(t)$ can be multivariate, in which case uncertainty about their values is expressed as covariance matrices ($\Sigma_a, \Sigma_p, \Sigma_y$) rather than scalar variances ($\sigma_a^2, \sigma_p^2, \sigma_y^2$). Similarly, the model prediction step has changed from multiplying the state variate with a constant scalar α to multiplication with a model matrix $\mathbf{F}$, with associated predictive uncertainty matrix Σ_f. [We could have extended the prediction step even further with an extra additive 'control-input' term representing the known impact of external driving variables $x(t)$ on the system state, which would still have left the KF analytically solvable.]

> KALMAN FILTERING (multivariate state):
>
> Kalman Gain: $\mathbf{K} = \Sigma_p \mathbf{H}^\top (\mathbf{H} \Sigma_p \mathbf{H}^\top + \Sigma_y)^{-1}$.
>
> Data assimilation: $p[z_a | \mu_p, \Sigma_p, y, \Sigma_y] = N[\mu_a, \Sigma_a]$,
>
> where $\mu_a = \mu_p + \mathbf{K}(y - \mathbf{H}\mu_p)$ and $\Sigma_a = (\mathbf{I} - \mathbf{K}\mathbf{H})\Sigma_p$.
>
> Prediction: $p[z_p | \mu_a, \Sigma_a] = N[\mu_p, \Sigma_p]$,
>
> where $\mu_p = \mathbf{F}\mu_a$ and $\Sigma_p = \mathbf{F}\Sigma_a\mathbf{F}^\top + \Sigma_f$.

$$(21.4)$$

The other major change in Eq. (21.4) compared to (21.3) is the *observational operator matrix* $\mathbf{H}$, which allows the observations to be a function of the states z rather than be the states themselves.

In KF, the product of observational operator and state estimate $\mathbf{H}z$ provides an estimate of the observations we expect to see when the system state is z. This is a very convenient component of the KF-procedure, because we often do not observe a system's state directly, but have to rely on measurement equipment that observes some derived property. An example would be where z consists of two components that differ in their temperature, but where we can only measure the overall average temperature. In that case, our observational operator would be a 1×2 matrix $\mathbf{H} = (0.5, 0.5)$ so that $\mathbf{H}z$ is the mean of the two temperatures $z[1]$ and $z[2]$. Time series models with such a separation between the true unobservable system state and observable properties that depend on the state are called *Hidden Markov Models* (HMM). So the KF is a HMM, although some people prefer to reserve the term HMM for time series models in which the system state space is discrete rather than continuous.

Let us now use the example of sensor-averaging to see a richer bivariate-state KF in action. Apart from z now being bivariate (and y measuring the average of the two), we use the same settings as for the simple univariate KF above. Our code for this matrix-version of Kalman Filtering is the following:

```
zp          <- Sp <- za <- Sa <- vector("list",n)
zp[[1]] <- c(0,0) ; Sp[[1]] <- Sf <- diag(1,2)
F           <- diag(0.9,2)
H           <- t( c(0.5,0.5) )
for(i in 1:n) {
   Ki         <- Sp[[i]] %*% t(H) %*% solve( H %*% Sp[[i]] %*% t(H) + Vy[i] )
   za[[i]]    <- zp[[i]] + Ki %*% ( y[i] - H %*% zp[[i]] )
   Sa[[i]]    <- ( diag(2) - Ki %*% H ) %*% Sp[[i]] ; if(i==n) break
   zp[[i+1]] <- F %*% za[[i]]
   Sp[[i+1]] <- F %*% Sa[[i]] %*% t(F) + Sf }
```

Note that our code above does not specify the observational uncertainties as covariance matrices because each $y(t)$ is a scalar as in our earlier example, so there is just a scalar observational variance at each time t. Also note that the code uses the R-data structure *lists* to collect the evolution of state estimates and uncertainty covariance matrices, which you can see from the use of the double square bracket notation [[ ]]. R-lists make it easy to address specific results. For example, $za[[30]]$ is the final state vector estimate (after the last data assimilation), and $Sa[[30]]$ is its uncertainty covariance matrix.

The results of this bivariate KF are shown in Fig. 21.3. Note that the figure shows the evolution of the mean and variance for our estimates of the *average* system state $(z(t) = (z_1(t) + z_2(t))/2)$, not its two individual components. The average state's variances are thus calculated as $V(t) = (V_1(t) + V_2(t) + 2\,Cov_{12}(t))/4$.

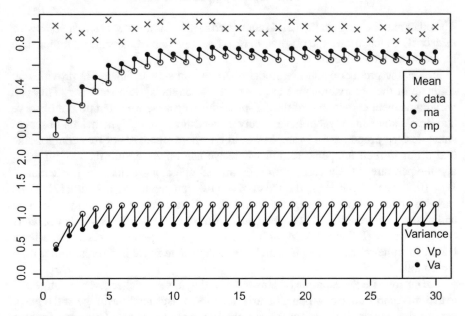

Fig. 21.3 Data assimilation using a bivariate KF. The top panel shows the observations and the alternating mean state estimates from model prediction (mp) and data assimilation (ma). The bottom panel shows the corresponding uncertainties. Observational uncertainty is not shown but has a constant variance equal to 3

When we compare the results of the bivariate KF in Fig. 21.3 with what we found before for the univariate KF 21.2, there are clear differences, which may be surprising. The state estimates and the variances are all lower in the bivariate case. But the same information from data was used in both cases, with the same observational uncertainties and the same parameter settings. We also used the same model f that at each time step reduced the last assimilation estimate by 10% at a predictive uncertainty of σ_f^2 equal to 1 resp. Σ_f equal to the bivariate unit matrix. So why did the model f seem to have a greater pull, relative to the data, in the bivariate case? The answer is that we have introduced extra information by applying our model to the two halves of the state separately. If z_p has a variance of 1, then $z_p/2$ has a variance of only 1/4. So if our knowledge of the system is the same in both cases, then the values on the diagonal of Σ_f should be higher than σ_f^2. How much higher is a question that we leave as an exercise.

21.3 Time Series, KF and Complex Dynamic Models

We have seen that the simple linear time series models of this chapter can generate a large variety of system behaviour. And their calculation is efficient, even when we include sequential state calibration. These advantages have made the KF a widely used data assimilation procedure. The computationally most demanding part of KF is the matrix inversion in the calculation of the 'Kalman Gain' $\mathbf{K}$ (see Eq. (21.4)). The algorithm can be faster if we first calculate the *singular value decomposition* (SVD) of $\mathbf{H}\Sigma_p\mathbf{H}^\top + \Sigma_y$ and then invert the SVD-components. We could code this in R as follows:

```
KalmanGain_SVD <- function(Sp,H,Vy) {
  Z     <- H %*% Sp %*% t(H) + Vy ; svd_Z <- svd(Z)
  inv_Z <- solve( t(svd_Z$v) ) %*% solve( diag(svd_Z$d) ) %*% solve( svd_Z$u )
  return( Sp %*% t(H) %*% inv_Z ) }
```

However, in the simple examples of this chapter, the SVD will not provide much computational advantage, nor will any other matrix decomposition.

The KF fits well in this book as its data assimilation step is Bayesian: the prior is the current predictive distribution for the state ($N[\mu_p, \Sigma_p]$), the likelihood is calculated from the difference of actual observations y and predicted observations ($\mathbf{H}\mu_p$) and the calculation of the posterior ($z_a|y \sim N[\mu_a, \Sigma_a]$) implicitly follows Bayes' Theorem without any need for MCMC because of the conjugate Gaussian prior-likelihood pair (see Chap. 5 for explanation of conjugacy).

However, the use of time series and sequential state corrections does have severe limitations. The first is that at each time step, there is no improvement to the model f. In data assimilation, we apply a correction to our state estimate in the direction of the observations, but we do not take away the need for that correction, which may be that our model is poorly parameterised. Bayesian calibration of model *parameters* is lacking. We do show elsewhere in this book that the KF could be used for Bayesian calibration of parameters (of linear models), but it would not be the obvious method for that (Chap. 8).

Further, in KF, the model structure is constrained to be a state-space model where f is a linear transformation of the state effected by the matrix $\mathbf{F}$. So how can we apply the KF when we already have a reliable but nonlinear predictive model? An obvious solution to this is to *linearize* the model around the last state estimate. This is done by replacing the nonlinear function f with the so-called Jacobian matrix $\mathbf{J}_f$ of all first order partial derivatives: $\mathbf{J}_f[i, j] = \partial f_i/\partial z_j$. You can then use $\mathbf{J}_f$ in the calculation of the predictive variance: $\Sigma_p = \mathbf{J}_f \Sigma_a \mathbf{J}_f^\top + \Sigma_f$. For the calculation of the predictive mean you can of course still use the nonlinear function f itself: $z_p = f(z_a)$ (instead of $z_p = \mathbf{F}z_a$). Likewise, the observational operator matrix $\mathbf{H}$ can be a linearization of a nonlinear function. A KF based on such linearization is called an *Extended Kalman Filter* (EKF). In practice, the EKF may perform adequately if the time step at which correction data arrive is small, so we do not need to predict far into the future. In other words, linearization is often acceptable as a *local* technique.

But what to do if our model f is so complex that we find it difficult to linearize, or when the time steps are too large? Our model may for example be a complex process-based model that predicts highly nonlinear system behaviour. We could then still apply the KF by creating an ensemble of multiple predictions from f (e.g. by sampling from its parameter distribution) and using the mean and variance of that set of predictions in the KF. That method is called *Ensemble Kalman Filtering* (EnKF) which is an example of the wider class of *particle filtering* methods. Of course, neither EKF nor EnKF address our concern that we want to improve our model structure and parameter values. Such data assimilation methods allow real-time correction for model discrepancy (Chap. 13) and poor parameterization without ever learning to make better predictions in the first place.

Exercises

1. Smoothness of a function represented by a GP. What would you need to change in the second code-chunk to produce more smooth curves?

2. Observational operator in KF. How would you specify the observational operator $\mathbf{H}$ for a KF where the system state is trivariate with only the second (middle) variate being observable such that the measurement signal is 80% of the variate's true value?

3. Bivariate vs. univariate KF. What very small change do you need to make in the R-code for bivariate KF (which led to Fig. 21.3) such that the results become identical to those of the univariate KF in Fig. 21.2)?

Chapter 22
Spatial Modelling and Scaling Error

22.1 Spatial Models

Much of spatial modelling is a generalization of the one-dimensional time series methods of the preceding chapter to two or more dimensions. Whereas time series methods discretize time, spatial models usually discretize space in the form of a raster of adjoining cells or a lattice of connected nodes. And Gaussian probability distributions feature equally prominently in both fields. An important class is that of *Gaussian Markov Random Fields* (GMRF) which generalize AR(1) models to multiple dimensions, and which includes common spatial model types such as *Conditional Autoregressive* (CAR) and *Simultaneous Autoregressive* (SAR) models, which are characterized by different covariance functions. In AR(1) models, the system state $z(t)$ is determined by the value of the state one time step earlier. In GMRF, which are defined on a lattice, the state $z(s)$ is determined by the values in the *neighborhood* of s, denoted as N_s. The neighborhood consists of the nodes that have a direct link (or 'edge') to s. So if we want to predict the value of $z(s)$ at some point in our space of interest, and the values of all $s' \in N_s$ are known, then nodes further away can provide no additional information. We say that every $z(s)$ is *conditionally independent* from nodes outside its neighborhood N_s. But note that if we were to write down the covariance matrix Σ for all points in our lattice, we would see a non-zero covariance for every pair of points irrespective of whether they are in each other's neighborhood or not. We already saw that in the AR(1) models of the preceding chapter where every $z(t)$ was calculated from its immediate predecessor $z(t-1)$, but where the covariance function of the equivalent GP extended to all times. However, the *precision matrix* $\Omega = \Sigma^{-1}$ of our GMRF, i.e. the inverse of the covariance matrix, only has non-zero values on its diagonal and for those $\Omega[i, j]$ where s_i and s_j are in each other's neighborhood. So the nodes and edges of the lattice structure tell us immediately where the non-zero values in Ω are. We actually already made use of this useful property of the precision matrix when we discussed graphical modelling (Chap. 15).

© Springer Nature Switzerland AG 2020
M. van Oijen, *Bayesian Compendium*,
https://doi.org/10.1007/978-3-030-55897-0_22

In the preceding chapter, we saw an example of an AR(1) time series model and an equivalent Gaussian Process (GP) model. The AR(1) was shown to be no more than a GP discretized on a sequence of times. Likewise a GMRF is a GP discretized on a lattice. But GP are in fact often directly employed in spatial modelling without discretization, with *geostatistics* as the main example. Let's investigate that further.

22.2   Geostatistics Using a GP

The standard problem in geostatistical modelling is that of spatial prediction. We want to predict the values of $z(s)$ in some region, but we will receive measurement values only for some locations in that area. In that situation the following statistical assumptions are generally made. We assume that locations which are near to each other will more likely have similar values of z than points further away. Part of that similarity can be captured by a deterministic function $m(s)$ of the spatial coordinates (and possibly of covariates $x(s)$) across the whole area. But some of the similarity from proximity will be due to mechanisms in nature that we have little knowledge of and which cannot be represented by simple deterministic functions. We model that unexplained spatial process by means of a covariance function (or 'kernel function') $C(s, s')$. Finally, we allow for the possibility that there is some non-spatially correlated noise that makes $z(s)$ differ even from points $z(s + \delta)$ very nearby. That noise term is for historical reasons (associated with the mining industry) called a *nugget effect* and it is modelled as a zero-mean Gaussian. Putting everything altogether, this statistical modelling approach to geostatistics can be written as (Diggle and Ribeiro 2007):

GEOSTATISTICS USING A GAUSSIAN PROCESS:

$$z(s) = m(s) + T(s) + \epsilon(s), \quad \text{where } T(s) \sim GP[0, C(s, s')] \text{ and } \epsilon(s) \sim N[0, \sigma_\epsilon^2].$$
$$(22.1)$$

This is a spatial linear Bayesian hierarchical model. It is of course almost the same statistical model as the one we studied in the chapter on model emulation (14). In that chapter our goal was to explore the output from a model f across its parameter space, whereas here we explore the state variable z across its physical space. The only difference is that we include the nugget-term $\epsilon(s)$ here. (But note that Andrianakis and Challenor (2012) showed that even for model emulation the use of a GP with a nugget can make the required matrix inversions more robust.)

Let's develop the geostatistics approach using a concrete example for a rectangular region whose coordinates are between 0 and 4 in the east-west direction and between 0 and 3 north-south. We use a negative-exponential covariance function (as we did in our emulation examples) and assign values to the hyperparameters. At first we leave out the nugget, but we shall include it later.

[Note that we solved a simpler version of this same problem in Chap. 15, where we had no nugget (like here) but also no true mean function, i.e. $m(s) = 0$, so no regression parameters needed to be estimated.]

```
x1 <- c(0,4,4)     ; x2  <- c(3,0,3) ; y <- c(0.7,3.1,2.2) ; ny <- length(y)
Vy <- 1.47         ; phi <- 1.9115
x <- cbind(x1,x2) ; dx <- as.matrix( dist(x) ) ; Sy <- exp( -dx/phi ) * Vy
```

Because our model has no nugget yet, we can apply the same parameter estimation function `GP.est` that we defined in the chapter on GP & emulation (Chap. 14). When supplying the three data points defined in the preceding code-chunk, this gives us the following parameter estimates:

Geostatistical parameter estimation using a GP:

$$\mu_{\beta|y} = \begin{bmatrix} 1.2859 \\ 0.4238 \\ -0.2328 \end{bmatrix} ; \quad \Sigma_{\beta|y} = \begin{bmatrix} 2.885 & -0.469 & -0.591 \\ -0.469 & 0.14 & 0.065 \\ -0.591 & 0.065 & 0.224 \end{bmatrix}. \tag{22.2}$$

Now let's see the predictive distribution for the point $y_0 = (0, 0)$. Deriving predictive distributions from a GP is also a problem we solved before in Chap. 14, and we use the R-function for prediction `GP.pred` that we developed there.

Geostatistical prediction for (0,0) using a GP:

$$p[y_0|y] = N[\,1.3181175,\ 3.7412996\,]. \tag{22.3}$$

22.3 Geostatistics Using `geoR`

Although in this book we prefer to use basic equations and basic R-code for our examples, to remove the mysteries of black-box modelling, it is of course occasionally convenient to use R-packages, and `geoR` is a very convenient package for Bayesian geostatistics. We have used it before in Chap. 14, where it was shown to produce the same parameter estimates and predictions as basic code. Here we show the code to set up our above model in `geoR`, and then use the package to produce maps of estimates and their uncertainties for the whole region.

```
xy      <- as.geodata( cbind( x, y ) )
xpred1  <- x0[1] ; xpred2 <- x0[2] ; xpred.1 <- expand.grid( xpred1, xpred2 )
model.1 <- model.control( cov.m="exponential",
                          trend.d=~coords, trend.l=~coords )
prior.1 <- prior.control( beta.prior="normal", beta=mb, beta.var.std=Sb/Vy,
                          sigmasq.prior="fixed", sigmasq=Vy,
                          phi.prior="fixed", phi=phi,
                          tausq.rel.prior="fixed",tausq.rel=0 )
out.1   <- output.control(messages=F)
geoR.1  <- krige.bayes( xy, loc=xpred.1, mod=model.1, pr=prior.1, out=out.1 )
```

Running this code and extracting the parameter estimates and variances from the `geoR.1` object produces the same results as we found in Equations (22.2) and (22.3), which you can verify by running the `geoR`-code yourself. So we are reassured that we can use the package to plot the geostatistical predictions and predictive uncertainties (standard deviations) for the whole region, including our prediction point (0,0) and

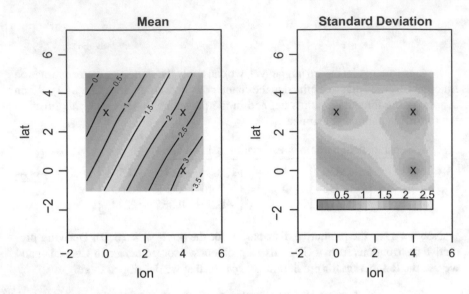

Fig. 22.1 Geostatistical prediction after calibration on three data points marked by crosses. GP without nugget. Left: predictive means; right: predictive standard deviations

the three training points. The results are shown in Fig. 22.1. The calibrated mean function $m(s)$ is shown on the left, showing contour lines mainly running from southwest to northeast. The right panel shows the predictive uncertainties (square roots of the variances). At the three measured points the uncertainty has become zero, while increasing with greater distance from those points.

22.4 Adding a Nugget

Adding a nugget of known variance is not difficult. In our basic equations, it amounts to adding a constant to all diagonal terms (= variances) of the spatial covariance matrix, without altering the off-diagonal covariances. And in geoR it is a matter of adding a non-zero relative nugget effect, denoted as tausq.rel in the package. We choose to add a nugget of 50% here. The only change required to the geoR-code to make this choice is in the definition of the prior distribution for parameters and hyperparameters:

```
prior.2 <- prior.control( beta.prior="normal", beta=mb, beta.var.std=Sb/Vy,
                          sigmasq.prior="fixed", sigmasq=Vy,
                          phi.prior="fixed", phi=phi,
                          tausq.rel.prior="fixed", tausq.rel=0.5 )
```

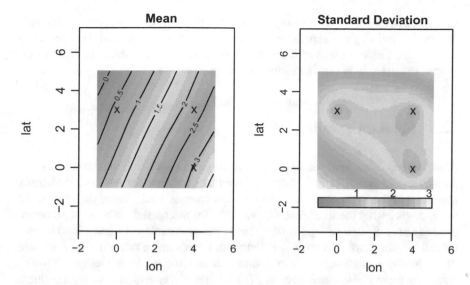

Fig. 22.2 Geostatistical prediction after calibration on three data points marked by crosses. GP with nugget. Left: predictive means; right: predictive standard deviations

Running this leads to Fig. 22.2. We see that the introduction of the nugget has had little effect on the mean surface, but predictive uncertainties have increased. For example, the predictive standard deviation for the point $(0, 0)$ increased from 1.934 to 2.242.

22.5   Estimating All GP-hyperparameters

So far we have only considered the case where the regression parameters β of the mean function of the GP are being estimated, with all other parameters considered to be known. Diggle et al. (2003) went a step further and showed an analytical solution for the case where β and a common scaling factor for Σ_β and Σ_y are estimated. Their analytical solution requires that the prior for the β is Gaussian, and for the variance-scaling factor is scaled inverse χ^2 (a form of inverse Gamma distribution). The marginal posterior distributions for β and for point prediction are then no longer Gaussian but more heavily-tailed t-distributions, reflecting the extra uncertainty that we have because of the unknown variances. Of course, a full Bayesian calibration would require all our uncertainties to be specified with prior distributions, but we quickly run out of analytical solutions if we try that. Diggle et al. therefore used MCMC to sample from the posterior distribution when also the correlation length ϕ was considered unknown, and that is how their package geoR operates when carrying out full Bayesian geostatistics. To speed up the computation, the package also requires that the domain of ϕ is discretized, so that the MCMC at least for that hyperparameter only needs to jump between a finite number of possible values. Note that such a full Bayesian spatial modelling approach would require an informative data set with observations at many more locations than the three we have in our simple example, so we leave this for the reader to explore.

22.6 Spatial Upscaling Error

As described in the last paragraph, the full Bayesian approach to geostatistics uses data to derive posterior probability distributions not just for the state variable at any point in the region but for all hyperparameters as well. We might be learning, for example, what the most probable value is for the spatial variance of some soil property $x(s)$. Spatial variability of $x(s)$ is of particular interest when x is an input to a model f, for example a process-based model. Let's make that more concrete with the following scenario that is very common in environmental modelling. We often want to use a model to make maps of $f(x)$ for some large region (country, continent, world), but have the problem that our model is slow to run. So we discretize space into many large grid cells, and for every grid cell we find the average value of x (which can be a scalar or vector), denoted as $E[x]$, and we calculate $f(E[x])$. Of course, we are then making an estimation error when f is nonlinear, because what we really should be calculating is $E[f(x)] = \int_{-\infty}^{+\infty} p[x] f(x) \, dx$, where $p[x]$ is the probability distribution that describes the variability of x in the cell. The erroneous assumption that $f(E[x]) = E[f(x)]$ is called the *fallacy of averages* and "is perhaps the most widespread statistical error in biology" (Welsh et al. 1988). It is a scaling error: we are applying a model that is designed to have *point-support* (i.e. use input from a single point in space) to a large area. Let's denote the upscaling error as $\Delta = f(E[x]) - E[f(x)]$. Figure 22.3 shows for scalar x that Δ is negative for convex functions, zero for linear functions, and positive for concave functions.

Fortunately, if we have a good estimate for the variance of x in the grid cell, then we can estimate the magnitude of the error from the Taylor expansion of f (Hines and Montgomery 1990):

$$E[f(x)] \approx f(E[x]) + \frac{1}{2}Var[x]f^{(2)}(E[x]) \implies$$
$$\widehat{\Delta} = -\frac{1}{2}Var[x]f^{(2)}(E[x]), \tag{22.4}$$

where $f^{(2)}(E[x])$ is the second derivative of f evaluated at the expectation value of x, and $Var[x]$ is the spatial variance of x. The formula tells us that (1) scaling error is expected to increase linearly with the variance of x, (2) its absolute value increases with the degree of nonlinearity of f (as measured by the magnitude of its second derivative), and (3) the sign of the scaling error (over- or underestimate) depends on the sign of the second derivative. That is quite a lot of information from such a simple equation!

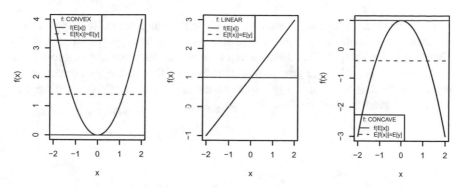

Fig. 22.3 Upscaling error for three functions f(x). The mean of f(x) in the interval $[-2,2]$ is calculated incorrectly as f applied to the mean of x (solid horizontal blue line) and correctly as the average of f(x) (dashed horizontal red line). Upscaling error f(E[x])–E[f(x)] is negative for convex functions, zero for linear functions, positive for concave functions

Let's take the very simple example of a quadratic function to show how the formula works: $f(x) = x^2$ with x distributed uniformly in the unit interval $[0,1]$. So $E[x] = 0.5$, $Var[x] = \frac{1}{12}$ and $f^{(2)}(E[x]) = 2$. Equation (22.4) then tells us to estimate the upscaling error as $\widehat{\Delta} = -\frac{1}{2}\frac{1}{12}2 = -\frac{1}{12}$. Now we need to verify that that is a good estimate. We find that $f(E[x]) = \frac{1}{2}^2 = \frac{1}{4}$ and $E[f(x)] = \int_0^1 x^2 dx = \frac{1}{3}$, so $\Delta = \frac{1}{4} - \frac{1}{3} = -\frac{1}{12}$ which confirms that Eq. (22.4) did give us the correct error estimate for this particular function. Briefly returning to Fig. 22.3, which showed that the sign of the scaling error depends on the type of nonlinearity of the function, we see that Eq. (22.4) indeed estimates a negative Δ for functions with a positive second derivative, i.e. convex functions.

Equation (22.4) is formulated for scalar x but it generalizes in a straightforward way to vectors (Van Oijen 2017):

$$\widehat{\Delta} = -\frac{1}{2}tr(\Sigma_x \nabla_f^2), \tag{22.5}$$

where Σ_x is the covariance matrix of x, ∇_f^2 is the *Hessian matrix* of second-order partial derivatives of $f(x)$, and tr denotes the trace (sum of diagonal elements) of the matrix product $\Sigma_x \nabla_f^2$. This error estimation formula is suprisingly powerful. For several combinations of function shape and input probability distribution, the formula is exact, $\widehat{\Delta} = \Delta$, while in many others the formula provides a good approximation (Van Oijen et al. 2017). We tested the formula with three existing nonlinear models for emissions of greenhouse gases (methane, ammonia, nitrous oxide) across the UK and found UK-average upscaling errors for the three models to be -12, -48 and -3%, in each case well estimated using the $\widehat{\Delta}$-formula. Of course, when the model f is not analytical, the Hessian matrix needs to be estimated numerically, and we provided R-code for the calculation of both the Hessian and $\widehat{\Delta}$ in the paper just mentioned.

Let's conclude. When we have used a fully Bayesian geostatistical approach to generate a spatial map of x, then we will also have acquired a posterior probability distribution for $Var[x]$. That uncertainty about spatial variance can be propagated forward using Eq. (22.4) to give us a probability distribution for the correction we need to make after we erroneously quantified the areal average of $f(x)$ as f applied to the areal average of x.

Chapter 23
Spatio-Temporal Modelling and Adaptive Sampling

23.1 Spatio-Temporal Modelling

The previous chapter showed the similarity of models for time series and for spatial variation. So we should not expect that *spatio-temporal modelling* adds any completely new algorithms. Spatio-temporal modelling aims to estimate the changes over time of a spatially distributed dynamic system. We could actually use pure time series models $z(t)$ or spatial models $z(s)$ to achieve that, by defining z as location or time, respectively. So $z(t)$ could represent an individual's changing location over time t, and $z(s)$ could be the time at which a wildfire reaches location s. But the term 'spatio-temporal modelling' usually refers to models where the state variable z changes as a function of both spatial and temporal coordinates, and is written as $z(s, t)$. Still, that does not require us to move to a new kind of model. We already saw that kernel-based approaches such as Gaussian Process modelling (GP) can be used for both spatial and temporal modelling, with the difference between the two mainly being one of dimensionality. A spatio-temporal system could certainly be modelled by means of a GP in three-dimensional space-time if we can define a sensible covariance function $C[(s, t), (s', t')]$. However, that is generally a difficult thing to do. It is very difficult to imagine what the proper covariance should be between our state variable z at location s and time t with the value of z at another place and time. Moroever, even if we could do so, this would lead to very large matrices that are difficult to invert and thus would cause computational problems for GP-estimation and -prediction. Therefore most spatio-temporal models employ *separable covariance functions* that can be decomposed as $C[(s, t), (s', t')] = C_s[s, s']C_t[t, t']$. The question of how similar we expect the system to be at nearby locations is thereby separated from the question of how we expect the system to evolve over time.

© Springer Nature Switzerland AG 2020
M. van Oijen, *Bayesian Compendium*,
https://doi.org/10.1007/978-3-030-55897-0_23

23.2   Adaptive Sampling

Separable spatio-temporal models can be very good tools to guide environmental
monitoring. For example, if we want to keep track of the level of air pollution in a
region, and only have a limited number of movable sensors available, then a spatio-
temporal model can help us deploy the sensors at those locations which would most
reduce our uncertainty about the regional average pollution level. Between measurements, we would be moving the sensors to new locations that minimize the model's
predictive variance. In other words, the model would help us with *adaptive sampling*
of the pollution levels across the region. We shall now look at a very simple example
of such spatio-temporal modelling, where our region of interest is a 7×7 spatial
grid, and where we follow the system over 8 time steps. So both space and time
are discretized. Our example is based on a very similar one provided by Wikle and
Royle (1999), who in turn based their model on the separable spatio-temporal model
of Huang and Cressie (1996) for melt-water prediction. The model uses an AR(1)
model for time and the Kalman Filter (KF) equations for updating the spatial distribution, but we have already seen in the last two chapters that both are equivalent to
GP-modelling. Initially, we set the temporal autoregression coefficient α to 0.9, and
the spatial autoregression coefficient ρ to 0.75. Note that the spatial KF-calculation
is for a 49-dimensional state variable z, because we have 49 different cells. We do
not estimate any parameters or hyperparameters here, we just examine how sensor
placement affects uncertainty about the state variable z.

   We focus in this very simple example on changes in the *variance* of the state
variable z, rather than on z itself. Given the simple model here (where all hyperparameters are assumed known), the variance is not informed by the data values y, only
by the data locations and the variance for measurement uncertainty Vy. In fact, we
arbitrarily choose the data values y to always be zero. What we want to minimize by
sensible repositioning of a single roving sensor is the *Average Predictive Variance*
(APV). So at each time step, our question is "whereto should we now move our sensor such that the average variance over all 49 grid cells is minimal?". We initialize
our spatio-temporal model with R-code that you are familiar with from the preceding
chapters:

```
coords    <- as.matrix( expand.grid(1:7,1:7) ) ; nz   <- dim(coords)[1]
tausq     <- 10 ; Vy <- diag( tausq, 1 ) ; ny <- 8 ; y <- 0
mb        <-  0 ; Va <- 80 ; rho <- 0.75
Rspatial  <- rho^as.matrix( dist(coords) )
zp        <- vector("list",ny)   ; Sp      <- vector("list",ny)
za        <- vector("list",ny+1) ; Sa      <- vector("list",ny+1)
za[[1]]   <- rep( mb, nz )        ; Sa[[1]] <- Va * Rspatial
APV       <- rep(NA,ny+1)         ; APV[1]  <- mean( diag( Sa[[1]] ) )
alpha     <- 0.9                  ; M <- diag(alpha,nz)
Vf        <- 20                   ; Q <- Vf * Rspatial
iSensor   <- rep(NA,ny)
```

   After this initialization, our spatio-temporal algorithm for adaptive sampling will
cycle through the following four steps:

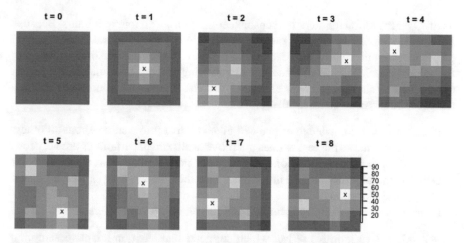

Fig. 23.1 Adaptive sampling by a single roving sensor in a region of 7×7 grid cells. Cell colours indicate predictive variance from low (white) to high (black). Current sensor position is marked by a cross

1. Forecast step: using the model to propagate `za[[i]],Sa[[i]]` to `zp[[i]],Sp[[i]]`.
2. Design step: selecting the sensor placement that gives lowest APV among all 49 possible locations.
3. Observation step: using the selected sensor location to measure y (this step can be skipped as long as we are only interested in the magnitude of predictive uncertainty).
4. Analysis step: using the sensor information (in particular its location) to update predictive variance.

We code this cycle in R as follows:

```
KalmanGain <- function(Sp,H,Vy){ Sp %*% t(H) %*% solve(H%*%Sp%*%t(H) + Vy) }
for (i in 1:ny) {
  zp[[i]] <- M %*% za[[i]] ; Sp[[i]] <- M %*% Sa[[i]] %*% t(M) + Q
  r_APV <- rep(NA,nz)
  for (r in 1:nz) {
    r_Hi      <- matrix(0,nrow=1,ncol=nz) ; r_Hi[r] <- 1
    r_Ki      <- KalmanGain( Sp[[i]], r_Hi, Vy )
    r_Sa      <- ( diag(nz) - r_Ki %*% r_Hi ) %*% Sp[[i]]
    r_APV[r]  <- mean( diag(r_Sa) ) }
  iSensor[i] <- match( min(r_APV), r_APV )
  Hi         <- matrix(0,nrow=1,ncol=nz) ; Hi[iSensor[i]] <- 1
  Ki         <- KalmanGain( Sp[[i]], Hi, Vy )
  za[[i+1]]  <- zp[[i]] + Ki %*% ( y - Hi %*% zp[[i]] )
  Sa[[i+1]]  <- ( diag(nz) - Ki %*% Hi ) %*% Sp[[i]]
  APV[i+1]   <- mean( diag( Sa[[i+1]] ) ) }
```

The results of this exercise are shown in Fig. 23.1. We see that the algorithm initially places the sensor in the exact centre, from where it reduces uncertainty as measured by the predictive variance in all directions because of the autoregressive spatial covariance function. The uncertainty is of course lowest at the sensor position itself. Over the following steps the sensor keeps exploring all four quadrants. Because this spatio-temporal system has memory (quantified by the temporal autoregression coefficient α), uncertainty stays reduced at locations where the sensor was placed in the recent past. So how would the sensor be moving around if the spatial correlation (ρ) or memory strength (α) were changed? You can explore that yourself by running the code, but here are some results:

- If we set $\rho = 1$, knowledge of one cell gives perfect information about all others at the same time, so the sensor does not move at all, and uncertainties are very low.
- If we set $\rho = 0$, knowledge of one cell gives no information at all about other cells, so the algorithm starts in a corner and then always moves to the nearest not-yet-used location.
- If we set $\alpha < 0.3$, memory is limited and the sensor stays at the central location.

All this makes intuitive sense, which suggests that such an adaptive sampling algorithm can be useful for managing monitoring campaigns.

23.3   Comments on Spatio-Temporal Modelling and Adaptive Sampling

Our adaptive sampling example can easily be extended to include parameter estimation, which would make the actual measurement values y informative. That can be done by extending the state vector with entries for the parameters. We already gave an example of using the KF for parameter estimation when we compared methods for Bayesian estimation of linear models (Chap. 8). We could even consider expanding the Bayesian learning during adaptive sampling to include learning which spatio-temporal *model* is to be preferred! For example the choice between different spatial covariance functions, where we start with a Bayes Factor equal to one before the sampling has commenced. That approach would be comparable to adaptive sampling algorithms used in MCMC.

Adaptive sampling could be used for the development of emulators too. That would mean optimising locations in parameter space (where to simulate next) rather than real space (where to measure next).

In our example we made the assumption that our quantity of interest was the average predictive variance. In practical problems, with complicated logistics and costs of sensor movement, maintenance and repair, the quantity of interest may deserve to be chosen more carefully. We could use Bayesian decision theory (Chap. 17) to assess the trade-off between costs (time and money) and gains (reduction of uncertainty).

For several more recent examples of Bayesian GP-assisted sampling algorithms, see Xu et al. (2016).

Chapter 24
What Next?

We are nearing the end of this book, and it is time to take stock. How far have we come and what can we do next?

I hope that this book has made clear why Bayesian methods are increasingly being used by scientists. The approach offers a consistent way of quantifying and reducing uncertainties, rigorously based on probability theory. The concepts are simple, consisting of no more than defining and combining priors and likelihoods. It provides us with a straightforward procedure for using data and assessing the reliability of models and model predictions.

The Bayesian outlook also helps us to see the forest for the trees in computational statistics. There is an enormous amount of confusing terminology, and we can choose from a very large number of different statistical methods with impressive-sounding names. But under the hood the differences are often small. Many statistical models are based on the same underlying mathematics and statistical assumptions, as we have mentioned in several chapters. Machine learning is just generalized regression, geostatistics and emulation are both applications of Gaussian processes, graphical modelling and hierarchical modelling are both tools for decomposing joint probability distributions. Kalman Filtering, which is commonly presented as a form of recurrent state estimation, can also be used for estimating parameter values.

Much of the applied statistics literature is about novel ways to linearize, Gaussianise ('normalise' is already taken) and discretize complexities away such that we can keep using standard techniques. It is all in the service of providing computationally efficient and sufficiently accurate estimates with well-quantified uncertainty. But even where methods are truly new, we can always try to understand them—and see how they are related—by asking the Bayesian questions: What is the statistical model? What is the stated or implied prior distribution? What is the likelihood function? How is Bayes' Theorem (explicitly or implicitly) used to find the posterior distribution? How is the posterior distribution used to assess the significance of the results?

© Springer Nature Switzerland AG 2020
M. van Oijen, *Bayesian Compendium*,
https://doi.org/10.1007/978-3-030-55897-0_24

I hope that seeing the family relationships between all kinds of different statistical methods has helped you get an overview of the field, and an understanding of the generality of Bayesian thinking. Jaynes used to refer to probability theory as "the logic of science", but it is of course everyone's logic. There is no difference in thinking between statisticians and scientists.

24.1 Some Crystal Ball Gazing

The following is my perspective on the future of Bayesian methodology, which, as the result of using a crystal ball, comes with considerable uncertainty. Some of the ideas were already mentioned by (Van Oijen 2017).

Bayesian methods are computationally demanding and their recent uptake by many modellers has only been made possible by the advent of fast computational methods for sampling from distributions, in particular MCMC. This has stimulated interest in the application of Bayesian methods to complex, slow models and to complex uncertainty assessments where multiple models are considered or where parameter uncertainty is defined hierarchically. But in such cases, computational demand becomes a bottleneck again, and we expect that the search for new, faster MCMC-algorithms and reliable approximation methods will continue. This will include attempts to develop better (faster and more reliable) model emulators, and there we may well see an increasing role of machine learning. A separate line of research will be efforts to design mechanistic rather than statistical emulators using ideas from the field of systems identification.

When we develop faster methods for process-based modelling (PBM), it will also become possible to embed them in Bayesian hierarchical models (BHM). That has so far not been possible, because of the computational demand. Bayesian calibration of a parameter-rich and slow process-based model is already hard, and allowing its parameters to vary in space or within a population makes it impossible. However, for scientists, a 'BHM-PBM' would constitute the holy grail as it would combine mechanistic understanding of the system (in the PBM) with knowledge of variability (in the BHM).

Another important issue that will remain is how to include the discrepancy between model behaviour and real system dynamics in the Bayesian analysis. We expect intensified methodological research on how to represent discrepancy and prior uncertainty about it. Ensemble modelling, which addresses model structural uncertainty too, is also likely to remain of key interest.

Besides *quantifying* model discrepancy, Bayesian methods will also be increasingly used to *reduce* discrepancy, i.e. improve model structure. Bayesian calibration, by reducing uncertainty about model parameter values, highlights the contribution of model structural error to model-data mismatch. Can supervised Bayesian machine learning methods help us design more realistic process-based models?

Uncertainties that have been quantified and reduced using Bayesian methods must be communicated to end-users of the modelling studies, but there is no standard approach for reporting uncertainties. However, research on uncertainty communication is continuing and graphical modelling may play an increasingly prominent role. Graphical models have, for example, been used to convey uncertainty about the impact of management decisions on the different services that a single ecosystem can provide (e.g. Smith et al. 2012).

These issues may acquire even more significance in the future with the expected increase in frequency and intensity of extreme climatic events, and their concomitant risks (Reyer 2015). Risk analysis is a form of uncertainty analysis and will therefore benefit from Bayesian methods in order to provide reliable estimates of risk, vulnerability and hazard probability. But risk analysis as practiced is often informal; it needs to be formulated more rigorously and embedded within Bayesian decision theory.

Finally, and in a very general sense, I believe that there will be a convergence or linking of methodology that we use for explanation (such as process-based models), prediction (such as machine learning) and uncertainty quantification (Bayes). The convergence will be stimulated by common goals, such as identifying causal pathways from an ever-increasing stream of data.

24.2 Further Reading

I hope that this book has stimulated your interest in Bayesian methods, and in probabilistic thinking more generally. If you want to read more, there are of course thousands of Bayesian research papers to study. There are also many good books, and I can only mention a few of them. The following three books have been written by and for scientists: *Probability theory: the logic of science* by Jaynes (2003), *Bayesian Logical Data Analysis for the Physical Sciences* by Gregory (2005) and *Data Analysis: A Bayesian Tutorial* by Sivia and Skilling (2006). The book by Jaynes is unique in its clarity and depth. It does not discuss many statistical models, nor any computational algorithms—it is a book about how to think. It demonstrates the strength of Bayesian thinking and how it resolves many paradoxes of the classical statistical literature. The books by Gregory and Sivia do discuss some concrete ways of analysing data. They focus on topics that are covered in the first half of this book, with Gregory using examples in the computer-algebra system Mathematica whereas Sivia stays with (very readable) mathematical exposition. A stand-out introductory text is *Statistical Rethinking* by McElreath (2016). He uses code-examples in R and Stan but the English-language explanations in particular form an impressively clear and jargon-free introduction to Bayes. This does come at the cost of requiring many pages of text to convey the ideas. *Bayesian Data Analysis* by Gelman et al. (2013) is the bible of Bayesian statisticians. It is not aimed at scientists, but it is where you may want to look up how different statistical models work, their underlying assumptions as well as their strengths and weaknesses.

24.3 Closing Words

The main message of this book is that Bayesian methods give you complete freedom
in data analysis and modelling. You can choose any probability distribution that
adequately represents your state of knowledge, not restricting yourself to Gaussians
for mathematical convenience. Bayes' Theorem always applies. And the advent of
sampling-based Bayesian calibration means that you can choose to work with realistic
nonlinear models, such as PBMs. You need not rely on the methods that statisticians
needed to invent, and the assumptions that they needed to make, when computing
power was limited and models needed to be solved by hand. I hope that you will go
out and enjoy the freedom that Bayesian methods coupled with computing power
give the modern scientist!

Appendix A
Notation and Abbreviations

Notation and Abbreviations

The two main conventions that we follow are:

1. Square brackets are used with probabilities, probability distributions, expectations and likelihoods: $p[\theta|y]$, $N[\mu, \sigma^2]$, $E[x]$, $L[\theta]$ etc.
2. Parentheses are used for functions, such as $f(x, \theta)$.

The most commonly used symbols are:

	Mathematics	R-code
Likelihood	$L[\theta]$, $p[y|\theta]$	L
Variance (scalar); Covariance matrix	Var[ ] or σ^2; Σ	V, S
Precision matrix	W	W
Covariates, predictor variables	x	x
Design matrix	X	X
Observations	y	y
True value or model output	z	z
Parameter (scalar or vector)	θ, β	b
Maximum likelihood estimate	θ^{MLE}	–
Prior mean and covariance matrix	μ_β, Σ_β	mb, Sb
Posterior mean and covariance matrix	$\mu_{\beta|y}$, $\Sigma_{\beta|y}$	mb_y, Sb_y
Covariance matrix for measurement uncertainty	Σ_y	Sy
Correlation length	ϕ	phi

© Springer Nature Switzerland AG 2020
M. van Oijen, *Bayesian Compendium*,
https://doi.org/10.1007/978-3-030-55897-0

Abbreviations

AR	Autoregressive time series model
A-R	Accept-Reject (or Rejection sampling)
BC	Bayesian Calibration
BDT	Bayesian Decision Theory
BHM	Bayesian Hierarchical Model
BMA	Bayesian Model Averaging
BMC	Bayesian Model Comparison
DAG	Directed Acyclic Graph
EXPOL5, EXPOL6	5- and 6-parameter version of the expolinear model
GBN	Gaussian Bayesian Network (or Gaussian Belief network)
geoR	R-package for Bayesian geostatistics
GLS	Generalized Least Squares
GM	Graphical Model
GP	Gaussian Process
JAGS	Just Another Gibbs Sampler (software for MCMC)
KF	Kalman Filter
MCMC	Markov Chain Monte Carlo
PBM	Process-Based Model
PRA	Probabilistic Risk Analysis

Appendix B
Mathematics for Modellers

This book is not very mathematical, but here are some elementary pointers.

How to read an equation

There are three ways to check equations:

1. The scientist's way: check dimensions,
2. The mathematician's way: replace variables with 0, 1 or ∞ (infinity),
3. Everyone's way: make a plot.

Let's take the example of the following equation:

$$p[\theta] = \frac{1}{\sigma\sqrt{2\pi}} \exp\left[-\frac{1}{2}\left(\frac{\theta - \mu}{\sigma}\right)^2\right].$$

It defines the function $p[\theta]$. Now we apply our three checks:

1. θ and μ must be in the same units, because they are subtracted from each other. And σ must be in the same units too, because exponentiated quantities (everything between the brackets of the exp()-function) must be dimensionless. That means that $p[\theta]$ has the inverse units of θ itself. That is as far as the dimensional analysis can take you for this equation: your physical knowledge of θ should now take over. [If you happen to know that the equation defines a probability density for θ, then you will be happy to see $p[\theta]$ having the inverse dimensions of θ.]
2. After various replacements with 0, 1 or ∞, you will know that:

   - the function goes to zero when θ or μ goes to plus or minus infinity,
   - if θ and μ are both replaced by the same constant, then the function reaches its maximum value,

© Springer Nature Switzerland AG 2020
M. van Oijen, *Bayesian Compendium*,
https://doi.org/10.1007/978-3-030-55897-0

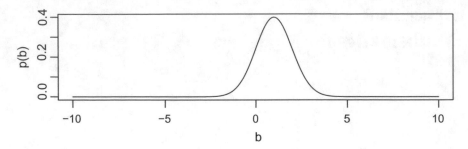

Fig. B.1 A function

- σ cannot be zero because that would give a division by zero.
- a very large value for σ would make $p[\theta]$ close to zero everywhere.

3. Plotting gives an even better idea of the shape of the function. Let's replace μ and σ by 1, and plot the function. Here's R-code and the result is in Fig. B.1

```
p <- function(b,m=1,s=1) {exp( -0.5 * ((b-m)/s)^2 ) / (s*sqrt(2*pi))}
b <- seq(-10,10,by=0.1) ; plot( b, p(b), type="l" )
```

The figure confirms what we already had found out about the behaviour of $p[\theta]$, and the R-function p() that we wrote gives us the opportunity to explore the function for different values of μ and σ.

Dimension Checking of Linear Algebra

There is some linear algebra in this book, because that makes it easy to work with multivariate quantities, and apply linear transormations to them. Take the following equation from Chap. 8:

$$\mu_{\beta|y} = \mu_\beta + \mathbf{K}(y - \mathbf{X}\mu_\beta).$$

Here $\mu_{\beta|y}$, μ_β and y are vectors and $\mathbf{K}$ and $\mathbf{X}$ are matrices. The equation defines how each element in μ_β is transformed into the corresponding element of $\mu_{\beta|y}$, so we are effectively writing multiple equations in one go.

Linear algebra comes with its own *mathematical* dimension check. Let's say that the dimensions of $\mathbf{K}$ are 2×3 (number of rows times number of columns: the rows are always given first). Knowing the dimensions of $\mathbf{K}$ means that the dimensions of all other quantities in the equation are determined too! This is because of how multiplication in linear algebra works. Say we have a vector or matrix A with dimensions $(r_A \times c_A)$, and a vector or matrix B that is $(r_B \times c_B)$. Now, we can only multiply them as $C = AB$ if $c_A = r_B$ and the dimensions of C will then be $(r_A \times c_B)$. If we now go back to our equation where $\mathbf{K}$ is 2×3, we find that the μ-vectors must be (2×1), the y-vector 3×1 and the matrix $\mathbf{X}$ (3×2), as you can verify for yourself.

Appendix C
Probability Theory for Modellers

Notation

- $p[x]$ = Probability of x. We use this notation not only for discrete distributions but also for continuous ones where we should strictly speak of the 'probability density' of x.
- $\neg x$ = Negation of x. For example, $p[\neg x]$ is the probability of x not being true.

Probability Distributions

A probability distribution is a non-negative function that sums to 1 (if it has a discrete domain) or integrates to 1 (if its domain is continuous). You need not rely on probability distributions that already have a name, and can make your own if you want.

Product Rule of Probability

$$p[A \cap B] = p[A]\, p[B|A] = p[B]\, p[A|B], \tag{C.1}$$

where $p[A \cap B]$ stands for the probability that both A and B are true.

Law of Total Probability

If $p[\theta]$ obeys a discrete distribution, then

© Springer Nature Switzerland AG 2020
M. van Oijen, *Bayesian Compendium*,
https://doi.org/10.1007/978-3-030-55897-0

$$p[y] = p[\theta]\,p[y|\theta] + p[\neg\theta]\,p[y|\neg\theta] = \sum_i p[\theta_i]\,p[y|\theta_i]. \qquad (C.2)$$

If $p[\theta]$ obeys a continuous distribution, then

$$p[y] = \int p[\theta]\,p[y|\theta]\,d\theta. \qquad (C.3)$$

Bayes' Theorem

$$p[\theta|y] = \frac{p[\theta]\,p[y|\theta]}{p[y]} \qquad (C.4)$$

Sequential Bayesian Updating

Bayes' Theorem can be used to calculate how a dataset changes our prior probability distribution for parameter(s) $p[\theta]$. Say that we receive a dataset $y1$ and have calculated the posterior distribution $p[\theta|y1]$. Now assume that later we receive another, independent dataset $y2$, and want to calculate a posterior distribution that incorporates the information from both datasets, so $p[\theta|y2, y1]$. We can obviously do that by starting again at the beginning, with $p[\theta]$, and feeding in both datasets at the same time into Bayes' Theorem, rather than just $y1$. But it may be more efficient if we take the posterior distribution from using the first dataset, i.e. $p[\theta|y1]$, and consider that to be the prior distribution for a second application of Bayes' Theorem. This is a valid approach as we can show:

$$\begin{aligned}
p[\theta|y2,\,y1] &= \frac{p[\theta|y1]\,p[y2|\theta,\,y1]}{p[y2|y1]} \\
&\propto p[\theta|y1]\,p[y2|\theta] \\
&\propto p[\theta]\,p[y1|\theta]\,p[y2|\theta].
\end{aligned} \qquad (C.5)$$

The simplification in the second line is possible because $y1$ and $y2$ are supposed to be independent. In the last line, we expanded using Bayes' Theorem as it was applied to the first dataset. Note that the likelihood calculation in the last line consists simply of multiplying the likelihood functions for the two datasets. That shows that the two-step procedure is valid and gives the same result as the one step procedure.

Gaussian Probability Distributions

If a scalar parameter θ has a Gaussian (or 'normal') distribution with mean μ and variance σ^2, we can write that succinctly as $\theta \sim N[\mu, \sigma^2]$. We can also give the formula of the probability distribution for θ:

$$p[\theta] = \frac{1}{\sigma\sqrt{2\pi}} \exp\left[-\frac{1}{2}\left(\frac{\theta - \mu}{\sigma}\right)^2\right]. \tag{C.6}$$

If a parameter vector θ (dimension $n \times 1$) has a multivariate Gaussian distribution with mean μ and covariance matrix Σ, we can write that succinctly as $\theta \sim N[\mu, \Sigma]$. The formula for the multivariate Gaussian is easiest written with matrix notation. We first rewrite the formula for the univariate Gaussian such that the similarity with the subsequent multivariate distribution will be most apparent.

- Univariate Gaussian distribution:

$$p[\theta] = (2\pi\sigma^2)^{-1/2} \exp\left[-\frac{1}{2}(\theta - \mu)(\sigma^2)^{-1}(\theta - \mu)\right]. \tag{C.6'}$$

- Multivariate Gaussian distribution:

$$p[\theta] = |2\pi\Sigma|^{-1/2} \exp\left[-\frac{1}{2}(\theta - \mu)^{\top}\Sigma^{-1}(\theta - \mu)\right]. \tag{C.7}$$

A nice property of the multivariate Gaussian is that conditional probabilities, where some of the variates are exactly known, can easily be calculated. In geostatistics, for example, we often set up a multivariate Gaussian $p[z] = N[\mu, \Sigma]$ for a vector z of dimension $n \times 1$, where one variate z_0 is unknown and the remaining $n - 1$ are known. A probabilistic prediction for the single unknown value is then written as $p[z_0|z_1]$, where z_1 is the known sub-vector of dimension $(n - 1 \times 1)$. Before writing out the general formula for a conditional multivariate Gaussian, we first define five symbols.

- $z = \begin{pmatrix} z_0 \\ z_1 \end{pmatrix}, \mu = \begin{pmatrix} \mu_0 \\ \mu_1 \end{pmatrix}, \Sigma = \begin{pmatrix} \Sigma_{00} & \Sigma_{01} \\ \Sigma_{10} & \Sigma_{11} \end{pmatrix},$
- $m_0 = \mu_0 + \Sigma_{01}\Sigma_{11}^{-1}(z_1 - \mu_1),$
- $V_0 = \Sigma_{00} - \Sigma_{01}\Sigma_{11}^{-1}\Sigma_{10},$

where $\{z_0, \mu_0, \Sigma_{00}, V_0\}$ are scalars with dimension (1×1), $\{z_1, \mu_1, \Sigma_{10}, \Sigma_{01}^{\top}\}$ are $(n - 1 \times 1)$ vectors, and Σ_{11} is an $(n - 1 \times n - 1)$ matrix. With these definitions, the conditional probability for the multivariate Gaussian is:

$$p[z_0|z_1] = |2\pi V_0|^{-1/2} \exp\left[-\frac{1}{2}(z_0 - m_0)^\top V_0^{-1}(z_0 - m_0)\right]. \qquad (C.8)$$

In fact, the formula is still correct if z_0 is not a scalar but multivariate, but we shall be using the partitioning into one unknown and $n - 1$ knowns more often.

Appendix D
R

R is a freely available computer language for statistical and scientific computing
that keeps developing. There is a large and active user community that produces
new functionality in the form of *packages* that can be downloaded and installed
when needed. Here we only introduce those parts of R that are used in this book. For
scientists who desire a comprehensive introduction to R, a great resource is the online
book *R for Data Science* by Hadley Wickham and Garrett Grolemund: https://r4ds.
had.co.nz. *RStudio* (https://rstudio.com) is an integrated development environment
that makes working with R a joy, and it facilitates literate programming by means of
Yihui Xie's *R Markdown* and *bookdown* (https://bookdown.org).

Basic R Commands

Variables can be scalars, vectors, matrices:

```
x    <- 2
y    <- c(1,2)
yrow <- matrix( c(1,2), nrow=1) ; ycol <- matrix( c(1,2), ncol=1)
z    <- matrix( 0:3, nrow=2)
```

Note the three different vector definitions. y is neither a row- nor a column-vector.
R will try to interpret its intended meaning from how it is used in the code. $yrow$ is a
row-vector and $ycol$ is a column-vector. We can check the dimensions of variables
using the length() and dim() functions. length() is defined for all variables,
dim() only for those where the numbers of rows and columns are known.

We can use the print() and cat() functions to show the values of variables
on our screen:

```
cat( length(x) , length(y) , length(yrow) , length(ycol) , length(Z) )
> 1 2 2 2 4
cat( dim(yrow) , " ; ", dim(ycol) , " ; ", dim(Z) )
> 1 2 ; 2 1 ; 2 2
```

© Springer Nature Switzerland AG 2020
M. van Oijen, *Bayesian Compendium*,
https://doi.org/10.1007/978-3-030-55897-0

Ordinary multiplication uses the '$*$' operator, matrix multiplication uses '$\% * \%$'.

```
xZ <- x * Z
Zy <- Z %*% y ; yZ <- y %*% Z ; Zyc <- Z %*% ycol ; yrZ <- yrow %*% Z
```

Diagonal matrices can be specified with the `diag` function:

```
I5a <- diag( 5 ) ; I5b <- diag( rep(1,5) ) ; X <- diag( 1:3 )
```

The inverse and transpose of matrices are calculated with the `solve` and `t` functions:

```
invZ <- solve(Z) ; trZ <- t(Z)
```

R has a *list* data structure, which can combine variables of any type. Elements from the list can be addressed using double square brackets, or using list-name + dollar-sign + element-name:

```
newlist <- list( distrib="Bivariate Gaussian", m=c(0,0), S=diag(2) )
newlist[[1]] ; newlist$m
> [1] "Bivariate Gaussian"
> [1] 0 0
```

R has many features for working with probability distributions. Probability densities are derived with functions `dnorm()`, `dunif()`, etc. Random numbers are generated with `rnorm()`, `runif()`, etc.

```
dx <- dnorm( x, mean=1, sd=0.5 ) ; nu <- runif( 50, min=0, max=100 )
```

You can define your own functions, say a quadratic:

```
myf <- function(x){x^2}
```

Basic plotting is done with the `plot`-command:

```
plot( 1:10, myf(1:10), main="My function", xlab="x", ylab="y" )
```

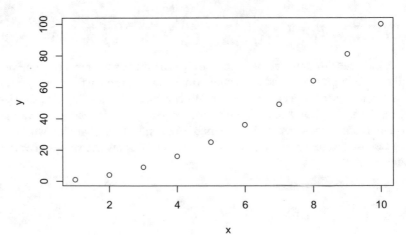

Appendix E
Bayesian Software

The code examples in this book were for the most part written in basic R, and they were often preceded by the relevant mathematical equations. Only on a few occasions were software packages with pre-defined functions used, and then only after examples in basic R had been given first. The idea was to avoid the black-box syndrome where you learn how to produce numbers but are not sure what you are doing, and cannot improve the algorithm. Despite the reliance on basic R, all code-chunks in this book were very short which confirmed the underlying simplicity of Bayesian methods. But you know all that now and may want to start using software that makes the implementation, execution and analysis of Bayesian methods just a little bit easier and faster. And free software exists that makes it easy to encode priors and likelihood functions and run an MCMC. These include, in historical order, the packages:

- BUGS (and more recently WinBUGS and OpenBUGS) (http://www.openbugs. net/w/FrontPage),
- JAGS (http://mcmc-jags.sourceforge.net),
- Stan (https://mc-stan.org/),
- Nimble (https://r-nimble.org/html_manual/cha-welcome-nimble.html).

These packages are currently mainly suited for hierarchical models with well-defined conditional probability distributions and not for complex PBMs. However, in recent years, two impressive pieces of software have been developed that facilitate the Bayesian calibration of PBMs using MCMC:

- BayesianTools (Hartig et al. 2017) can run different MCMC algorithms, and it has been incorporated in …
- PEcAn (Dietze et al. 2013) which allows the user to choose from a suite of different PBMs and driver conditions.

© Springer Nature Switzerland AG 2020
M. van Oijen, *Bayesian Compendium*,
https://doi.org/10.1007/978-3-030-55897-0

But if you have any specific statistical model or computational tool in mind, and you wonder whether any R-packages exist for that, your first port of call should be the CRAN repository of R-software, and in particular its 'Task View for Bayesian Inference': https://cran.r-project.org/web/views/Bayesian.html. There is a wealth of material there, including a link to `geoR` which we used in Chaps. 14 and 22.

Appendix F
Solutions to Exercises

Chapter 1

1. Yes, your distributions would converge.
2. Strictly, $p[\theta|y]$ and $p[\theta]$ are probability densities, not probability distributions. However, Bayes' Theorem is valid for any choice of θ and y, so distributions are implied.
3. The question is answered in the chapter.
4. Yes.

Chapter 5

1. Inference from the analytical formula. If $n = 0$ then the posterior is the same as the prior: $\mu_1 = \mu_0$ and $\sigma_1 = \sigma_0$. If $n \to \infty$ then the prior becomes irrelevant and the uncertainty disappears: $\mu_1 = \bar{y}$ and $\sigma_1 \to 0$.
2. Explore the A-R algorithm. This requires self-study.

Chapter 6

1. Metropolis for sampling from the posterior. The R-function for Metropolis that we defined in this chapter (`Metropolis(p,b0,SProp,ni)`) can be called with the product of prior and likelihood as its first argument, because all we need to provide that function with is a quantity that is proportional to our desired posterior distribution, and Bayes' Theorem tells us that the product of prior and likelihood fits the bill.

© Springer Nature Switzerland AG 2020
M. van Oijen, *Bayesian Compendium*,
https://doi.org/10.1007/978-3-030-55897-0

Chapter 7

1. Identifiability of parameters. The posterior parameter uncertainties will be much larger for the model with redundancy. And even if the number of data points would go to infinity, the uncertainty in the two slope parameters of the second model would remain large. But the uncertainty about their *sum* would approach zero. So both models (2- and 3-parameter versions) produce equally reliable output predictions after calibration. We conclude that parameter identifiability is generally a non-issue, unless for some reason you are more interested in the values of your model's parameters than in the model's predictions. Of course, there are other good reasons for removing redundant parameters, primarily that of keeping the model simple.

2. Convergence and posterior. The MCMCs will only behave identically if parameter settings for the prior and proposal are carefully chosen. For the first parameter of both models (their intercepts $\beta_1^{(1)}$ resp. $\beta_1^{(2)}$), you can use the same settings for prior and proposal. But to ensure that the MCMCs explore parameter space equally efficiently for the *slope parameter* of the first model ($\beta_2^{(1)}$) as for the *sum of the two slope parameters* ($\beta_2^{(2)} + \beta_3^{(2)}$) of the second model, the settings need to be different. The *sum* of the prior means, of the prior variances, and of the proposal variances for $\beta_2^{(2)}$ and $\beta_3^{(2)}$ should equal the corresponding values for $\beta_2^{(1)}$. Only then will the MCMCs produce the same posterior distribution for $\beta_2^{(1)}$ as for $\beta_2^{(2)} + \beta_3^{(2)}$, and thus also the same posterior predictive distribution for model output.

3. Which MCMC produced the following posterior samples? Panel A = MCMC 1; B = 6; C = 2 ; D = 5; E = 3; F = 4.

Chapter 8

1. Prior uncertainty. The posterior estimates for β become smaller, moving slightly in the direction of the prior mean of (0,0).

2. Kalman Filtering (KF) without intercept uncertainty. By setting $\Sigma_\beta[1, 1]$ to zero, you have effectively removed all uncertainty about the intercept: you fixed its value at the value that we specified in the chapter: zero. That changed the problem from estimating a two-parameter straight line to fitting a one-parameter line through the origin. The posterior uncertainty for the intercept remains zero (as it was in your modified prior), and the uncertainty for the slope is reduced because we now only needed to fit one parameter. Whereas very high or low values for the slope could previously be compensated for by low resp. high values of the intercept, that is no longer possible when we keep the intercept fixed.

```
K     <- KalmanGain( Sb*diag(c(0,1)), X, Sy )
Sb_y <- ( diag(nb) - K %*% X ) %*% Sb*diag(c(0,1))
mb_y <- mb + K %*% ( y - X %*% mb )
print(round(mb_y,3)) ; print(round(Sb_y,3))
```

Chapter 14

1. GP with small correlation length. If you set $\phi = 0$, you will have carried out the linear regression with geoR that was announced in Chap. 8. Your results will also resemble those of the last panel of Fig. 14.1, which by neglecting spatial correlation is effectively linear regression.

Chapter 15

1. Bayes' Theorem as a DAG. $[\theta] \rightarrow [y]$ is the DAG for us, because $p[y|\theta]$ is easier to specify than $p[\theta|y]$. Of course, Bayes' Theorem will help us to track back from y to θ when measurements of y become available.

2. Covariance matrices (after MacKay). Convert the covariance matrices M1 .. M4 to precision matrices using R's solve() function, and then apply the VcondR() function, defined in the chapter, to see which R-matrices have none-zero entries in the required positions [2,1], [3,1], [3,2]. In short, type VcondR(solve(M1)) etc. The answer to the question is: M3 and M4.

Chapter 16

1. Taking the H out of BHM. Yes, the BHM will collapse to a non-hierarchical model if we make the priors for hyperparameter-variances extremely narrow.

Chapter 20

1. The neural network of Fig. 20.2. Here $n_p = n_h(n_x + 1) + n_y(n_h + 1) = 3 \times 3 + 2 \times 4 = 17$. For this small network, you could also see this directly from the figure: 12 arrow-weights plus 5 neuron- and output-biases.

2. 1-1-1 neural networks. Choose identity activation functions, biases equal to zero, and weights such that their product w1 * w2 = 2. For example: fNN.11(x,2,0,1,0,identity) or fNN.11(x,1,0,2,0,identity).

Chapter 21

1. Smoothness of a function respresented by a GP. Choose a larger value for `phi`, e.g. 100 or 1000.

2. Observational operator in KF. **H** = (0, 0.8, 0).

3. Bivariate vs. univariate KF. To make the bivariate KF equivalent to the univariate one, double the predictive uncertainty variances, i.e. set `Sf <- diag(2,2)`.

References

Andrianakis, I., and P.G. Challenor. 2012. The effect of the nugget on Gaussian process emulators of computer models. *Computational Statistics & Data Analysis* 56: 4215–4228. https://doi.org/10.1016/j.csda.2012.04.020.

Andrianakis, I., I.R. Vernon, N. McCreesh, T.J. McKinley, J.E. Oakley, R.N. Nsubuga, M. Goldstein, and R.G. White. 2015. Bayesian history matching of complex infectious disease models using emulation: A tutorial and a case study on HIV in uganda. *PLoS Computational Biology* 11: e1003968. https://doi.org/10.1371/journal.pcbi.1003968.

Andrieu, C., and J. Thoms. 2008. A tutorial on adaptive MCMC. *Statistics Computational* 18: 343–373. https://doi.org/10.1007/s11222-008-9110-y.

Bakka, H., H. Rue, G.-A. Fuglstad, A. Riebler, D. Bolin, E. Krainski, D. Simpson, and F. Lindgren. 2018. Spatial modelling with R-INLA: A review. arXiv:1802.06350 [stat].

Bayes, T. 1763. An essay towards solving a problem in the doctrine of chances. *Philosophical Transactions* 53: 370–418. https://doi.org/10.1098/rstl.1763.0053.

Berger, J.O. 1985. *Statistical decision theory and bayesian analysis*, 2nd ed. New York: Springer Series in Statistics. Springer.

Blangiardo, M., M. Cameletti, G. Baio, and H. Rue. 2013. Spatial and spatio-temporal models with R-INLA. *Spatial and Spatio-temporal Epidemiology* 4: 33–49. https://doi.org/10.1016/j.sste.2012.12.001.

Bolker, B.M. 2008. *Ecological models and data in R*. Princeton University Press.

Brynjarsdóttir, J., and A. O'Hagan. 2014. Learning about physical parameters: The importance of model discrepancy. *Inverse Problems* 30: 114007. https://doi.org/10.1088/0266-5611/30/11/114007.

Burnham, K.P., and D.R. Anderson. 2002. Model selection and multimodel inference: A practical information-theoretic approach. Springer.

Cameron, D.R., M. Van Oijen, C. Werner, K. Butterbach-Bahl, R. Grote, E. Haas, G.B.M. Heuvelink, R. Kiese, J. Kros, M. Kuhnert, A. Leip, G.J. Reinds, H.I. Reuter, M.J. Schelhaas, W. De Vries, and J. Yeluripati. 2013. Environmental change impacts on the C- and N-cycle of European forests: A model comparison study. *Biogeosciences* 10: 1751–1773. https://doi.org/10.5194/bg-10-1751-2013.

Carbajal, J.P., J.P. Leitão, C. Albert, and J. Rieckermann. 2017. Appraisal of data-driven and mechanistic emulators of nonlinear simulators: The case of hydrodynamic urban drainage models. *Environmental Modelling & Software* 92: 17–27. https://doi.org/10.1016/j.envsoft.2017.02.006.

© Springer Nature Switzerland AG 2020
M. van Oijen, *Bayesian Compendium*,
https://doi.org/10.1007/978-3-030-55897-0

Castillo, E., J.M. Menéndez, and S. Sánchez-Cambronero. 2008. Predicting traffic flow using Bayesian networks. *Transportation Research Part B: Methodological* 42: 482–509. https://doi.org/10.1016/j.trb.2007.10.003.

Chamberlin, T.C. 1890. The method of multiple working hypotheses. *Science* 15: 92–96.

Chandler, R.E. 2013. Exploiting strength, discounting weakness: Combining information from multiple climate simulators. *Philosophical Transactions of the Royal Society A: Mathematical, Physical and Engineering Sciences* 371: 20120388–20120388. https://doi.org/10.1098/rsta.2012.0388.

Chipman, H.A., E.I. George, and R.E. McCulloch. 2010. BART: Bayesian additive regression trees. *The Annals of Applied Statistics* 4: 266–298. https://doi.org/10.1214/09-AOAS285.

Clark, J.S., L. Iverson, C.W. Woodall, C.D. Allen, D.M. Bell, D.C. Bragg, A.W. D'Amato, F.W. Davis, M.H. Hersh, I. Ibanez, S.T. Jackson, S. Matthews, N. Pederson, M. Peters, M.W. Schwartz, K.M. Waring, and N.E. Zimmermann. 2016. The impacts of increasing drought on forest dynamics, structure, and biodiversity in the United States. *Global Change Biology* 22: 2329–2352. https://doi.org/10.1111/gcb.13160.

Conti, S., and A. O'Hagan. 2010. Bayesian emulation of complex multi-output and dynamic computer models. *Journal of Statistical Planning and Inference* 140: 640–651. https://doi.org/10.1016/j.jspi.2009.08.006.

Cressie, N., C.A. Calder, J.S. Clark, J.M.V. Hoef, and C.K. Wikle. 2009. Accounting for uncertainty in ecological analysis: The strengths and limitations of hierarchical statistical modeling. *Ecological Applications* 19: 553–570.

Cressie, N., and C.K. Wikle. 2011. *Statistics for spatio-temporal data*. Wiley.

Dietze, M.C., D.S. Lebauer, and R. Kooper. 2013. On improving the communication between models and data. *Plant Cell Environ* 36: 1575–1585. https://doi.org/10.1111/pce.12043.

Dietze, M.C., M.S. Wolosin, and J.S. Clark. 2008. Capturing diversity and interspecific variability in allometries: A hierarchical approach. *Forest Ecology and Management* 256: 1939–1948. https://doi.org/10.1016/j.foreco.2008.07.034.

Diggle, P.J., P.J. Ribeiro Jr., and O.F. Christensen. 2003. An introduction to model-based geostatistics. *Spatial Statistics and Computational Methods. Springer* 43–86.

Diggle, P., and P.J. Ribeiro. 2007. *Model-based geostatistics*. Springer.

Drignei, D. 2017. An estimation algorithm for fast kriging surrogates of computer models with unstructured multiple outputs. *Computer Methods in Applied Mechanics and Engineering* 321: 35–45. https://doi.org/10.1016/j.cma.2017.04.001.

DuMouchel, W., and S.-L. Normand. 2000. Computer-modeling and graphical strategies for meta-analysis. *Meta-analysis in Medicine and Health Policy* 127–178.

Fu, Y.H., and M., Campioli, M. Van Oijen, G. Deckmyn, and I.A. Janssens, 2012. Bayesian comparison of six different temperature-based budburst models for four temperate tree species. *Ecological Modelling* 230: 92–100. https://doi.org/10.1016/j.ecolmodel.2012.01.010.

Gabry, J., D. Simpson, A. Vehtari, M. Betancourt, and A. Gelman. 2019. Visualization in Bayesian workflow. *Journal of the Royal Statistical Society A* 182: 389–402. https://doi.org/10.1111/rssa.12378.

Gelman, A., J.B. Carlin, H.S. Stern, D.B. Dunson, A. Vehtari, and D.B. Rubin. 2013. *Bayesian data analysis, 3*, edition ed. Boca Raton: Chapman and Hall/CRC.

Gelman, A., and J. Hill. 2006. *Data analysis using regression and multilevel/hierarchical models*. Cambridge University Press.

Gelman, A., J. Hwang, and A. Vehtari. 2014. Understanding predictive information criteria for Bayesian models. *Statistics and Computing* 24: 997–1016. https://doi.org/10.1007/s11222-013-9416-2.

Gilks, W.R., S. Richardson, and D. Spiegelhalter. 1995. *Markov chain monte carlo in practice*. CRC Press.

Goldstein, M., 2015. Bayes linear analysis, in: Wiley StatsRef: Statistics reference online. *American Cancer Society*, 1–7. https://doi.org/10.1002/9781118445112.stat00225.pub2.

Goldstein, M., M.J. Bayarri, J.O. Berger, A.P. Dawid, D. Heckerman, A.F.M. Smith, and M. West. 2010. External Bayesian analysis for computer simulators. *Bayesian Statistics* 9: 17.

Goldstein, M., and J. Rougier. 2009. Reified Bayesian modelling and inference for physical systems. *Journal of Statistical Planning and Inference* 139: 1221–1239. https://doi.org/10.1016/j.jspi. 2008.07.019.

Gonzalez-Redin, J., S. Luque, L. Poggio, R. Smith, and A. Gimona. 2016. Spatial Bayesian belief networks as a planning decision tool for mapping ecosystem services trade-offs on forested landscapes. *Environmental Research, The Provision of Ecosystem Services in Response to Global Change* 144: 15–26. https://doi.org/10.1016/j.envres.2015.11.009.

Goudriaan, J., and J.L. Monteith. 1990. A mathematical function for crop growth based on light interception and leaf area expansion. *Annals of Botany* 66: 695–701.

Green, P.J. 1995. Reversible jump Markov chain Monte Carlo computation and Bayesian model determination. *Biometrika* 82: 711–732. https://doi.org/10.1093/biomet/82.4.711.

Gregory, P. 2005. Bayesian logical data analysis for the physical sciences: A comparative approach with mathematica. *Cambridge University Press, Cambridge*. https://doi.org/10.1017/CBO9780511791277.

Haario, H., M. Laine, A. Mira, and E. Saksman. 2006. DRAM: Efficient adaptive MCMC. *Statistics and Computing* 16: 339–354. https://doi.org/10.1007/s11222-006-9438-0.

Hartig, F., J. Dyke, T. Hickler, S.I. Higgins, R.B. O'Hara, S. Scheiter, and A. Huth. 2012. Connecting dynamic vegetation models to data – an inverse perspective. *Journal of Biogeography* 39: 2240–2252. https://doi.org/10.1111/j.1365-2699.2012.02745.x.

Hartig, F., F. Minunno, S. Paul, and D. Cameron. 2017. Package 'BayesianTools'.

Hauser, T., A. Keats, and L. Tarasov. 2012. Artificial neural network assisted Bayesian calibration of climate models. *Climate Dynamics* 39: 137–154. https://doi.org/10.1007/s00382-011-1168-0.

Hickler, T., A. Rammig, and C. Werner. 2015. Modelling CO2 impacts on forest productivity. *Current Forestry Reports* 1: 69–80. https://doi.org/10.1007/s40725-015-0014-8.

Higdon, D., J. Gattiker, B. Williams, and M. Rightley. 2008. Computer model calibration using high-dimensional output. *Journal of the American Statistical Association* 103: 570–583. https://doi.org/10.1198/016214507000000888.

Hines, W., and D. Montgomery. 1990. *Probability and statistics in engineering and management science*, 3rd ed. New York: Wiley.

Hjelkrem, A.-G.R., M. Höglind, M. van Oijen, J. Schellberg, T. Gaiser, and F. Ewert. 2017. Sensitivity analysis and Bayesian calibration for testing robustness of the BASGRA model in different environments. *Ecological Modelling* 359: 80–91. https://doi.org/10.1016/j.ecolmodel.2017.05.015.

Hojsgaard, S., D. Edwards, S. Lauritzen. 2012. Graphical Models with R.

Höglind, M., M. Van Oijen, D. Cameron, and T. Persson. 2016. Process-based simulation of growth and overwintering of grassland using the BASGRA model. *Ecological Modelling* 335: 1–15. https://doi.org/10.1016/j.ecolmodel.2016.04.024.

Huang, H.-C., and N. Cressie. 1996. Spatio-temporal prediction of snow water equivalent using the Kalman filter. *Computational Statistics & Data Analysis* 22: 159–175. https://doi.org/10.1016/0167-9473(95)00047-X.

Hyvönen, R., G.I. Ågren, S. Linder, T. Persson, M.F. Cotrufo, A. Ekblad, M. Freeman, A. Grelle, I.A. Janssens, P.G. Jarvis, S. Kellomäki, A. Lindroth, D. Loustau, T. Lundmark, R.J. Norby, R. Oren, K. Pilegaard, M.G. Ryan, B.D. Sigurdsson, M. Strömgren, M. van Oijen, and G. Wallin. 2007. The likely impact of elevated [CO2], nitrogen deposition, increased temperature and management on carbon sequestration in temperate and boreal forest ecosystems: A literature review. *New Phytologist* 173: 463–480. https://doi.org/10.1111/j.1469-8137.2007.01967.x.

James, G., D. Witten, T. Hastie, and R. Tibshirani. 2013. An introduction to statistical learning, springer texts in statistics. *Springer, New York, NY*. https://doi.org/10.1007/978-1-4614-7138-7.

Jaynes, E.T. 2003. *Probability theory: The logic of science*. Cambridge University Press.

Johnson, M.O., D. Galbraith, M. Gloor, H. De Deurwaerder, M. Guimberteau, A. Rammig, K. Thonicke, H. Verbeeck, C. von Randow, A. Monteagudo, O.L. Phillips, R.J.W. Brienen, T.R. Feldpausch, G. Lopez Gonzalez, S. Fauset, C.A. Quesada, B. Christoffersen, P. Ciais, G. Sampaio, B. Kruijt, P. Meir, P. Moorcroft, K. Zhang, E. Alvarez-Davila, A. Alves de Oliveira, I. Amaral, A. Andrade, L.E.O.C. Aragao, A. Araujo-Murakami, E.J.M.M. Arets, L. Arroyo, G.A. Aymard, C. Baraloto, J. Barroso, D. Bonal, R. Boot, J. Camargo, J. Chave, A. Cogollo, F. Cornejo Valverde, A.C. Lola da Costa, A. Di Fiore, L. Ferreira, N. Higuchi, E.N. Honorio, T.J. Killeen, S.G. Laurance, W.F. Laurance, J. Licona, T. Lovejoy, Y. Malhi, B. Marimon, B.H. Marimon, D.C.L. Matos, C. Mendoza, D.A. Neill, G. Pardo, M. Peña-Claros, N.C.A. Pitman, L. Poorter, A. Prieto, H. Ramirez-Angulo, A. Roopsind, A. Rudas, R.P. Salomao, M. Silveira, J. Stropp, H. ter Steege, J. Terborgh, R. Thomas, M. Toledo, A. Torres-Lezama, G.M.F. van der Heijden, R. Vasquez, I.C. Guimarães Vieira, E. Vilanova, V.A. Vos, and T.R. Baker. 2016. Variation in stem mortality rates determines patterns of above-ground biomass in Amazonian forests: Implications for dynamic global vegetation models. *Global Change Biology* 22: 3996–4013. https://doi.org/10.1111/gcb.13315.

Kapelner, A., and J. Bleich. 2016. bartMachine: Machine learning with bayesian additive regression trees. *Journal of Statistical Software* 70: 40. https://doi.org/10.18637/jss.v070.i04.

Kass, R.E., and A.E. Raftery. 1995. Bayes factors. *Journal of the American Statistical Association* 90: 773–795.

Kavetski, D., G. Kuczera, and S.W. Franks. 2006. Bayesian analysis of input uncertainty in hydrological modeling: 1. Theory. *Water Resources Research* 42: https://doi.org/10.1029/2005WR004368.

Kennedy, M.C., and A. O'Hagan. 2001. Bayesian calibration of computer models. *Journal of the Royal Statistical Society: Series B (Statistical Methodology)* 63: 425–464.

Kobayashi, K., and M.U. Salam. 2000. Comparing simulated and measured values using mean squared deviation and its components. *Agronomy Journal* 92: 345–352. https://doi.org/10.2134/agronj2000.922345x.

Kuhn, M. 2008. Building Predictive models in R using the caret package. *Journal of Statistical Software* 28: 1–26. https://doi.org/10.18637/jss.v028.i05.

Leeds, W.B., C.K. Wikle, J. Fiechter, J. Brown, and R.F. Milliff. 2013. Modeling 3-D spatio-temporal biogeochemical processes with a forest of 1-D statistical emulators: Modeling 3-D processes with a forest of 1-D emulators. *Environmetrics* 24: 1–12. https://doi.org/10.1002/env.2187.

Levy, P.E., N. Cowan, M. van Oijen, D. Famulari, J. Drewer, and U. Skiba. 2017. Estimation of cumulative fluxes of nitrous oxide: Uncertainty in temporal upscaling and emission factors: Estimation of cumulative fluxes of nitrous oxide. *European Journal of Soil Science* 68: 400–411. https://doi.org/10.1111/ejss.12432.

Levy, P.E., R. Wendler, M. Van Oijen, M.G. Cannell, and P. Millard. 2005. The effect of nitrogen enrichment on the carbon sink in coniferous forests: Uncertainty and sensitivity analyses of three ecosystem models. *Water, air, & soil pollution: Focus* 4: 67–74.

Lindley, D., and A. Smith. 1972. Bayes estimates for the linear model. *Journal of the Royal Statistical Society* 34: 1–41.

Lindley, D.V. 1991. *Making decisions*, 2 edition. ed. Wiley, London; New York.

Lintusaari, J., M.U. Gutmann, R. Dutta, S. Kaski, and J. Corander. 2017. Fundamentals and Recent Developments in Approximate Bayesian Computation. *Systematic Biology* 66: e66–e82. https://doi.org/10.1093/sysbio/syw077.

Loeppky, J.L., J. Sacks, and W.J. Welch. 2009. Choosing the sample size of a computer experiment: A practical guide. *Technometrics* 51: 366–376. https://doi.org/10.1198/TECH.2009.08040.

MacKay, D.J. 2003. *Information theory, inference and learning algorithms*. Cambridge University Press.

MacKay, D.J.C., 1995. *Bayesian methods for neural networks: Theory and applications*.

MacKay, D.J.C. 1992. Bayesian interpolation. *Neural Computation* 4: 415–447.

Mäkelä, A., M. del Río, J. Hynynen, M.J. Hawkins, C. Reyer, P. Soares, M. van Oijen, and M. Tomé. 2012. Using stand-scale forest models for estimating indicators of sustainable forest management. *Forest Ecology and Management* 285: 164–178. https://doi.org/10.1016/j.foreco.2012.07.041.

McElreath, R. 2016. *Statistical rethinking: A bayesian course with R examples*.

Medlyn, B.E., S. Zaehle, M.G. De Kauwe, A.P. Walker, M.C. Dietze, P.J. Hanson, T. Hickler, A.K. Jain, Y. Luo, W. Parton, I.C. Prentice, P.E. Thornton, S. Wang, Y.-P. Wang, E. Weng, C.M. Iversen, H.R. McCarthy, J.M. Warren, R. Oren, and R.J. Norby. 2015. Using ecosystem experiments to improve vegetation models. *Nature Climate Change* 5: 528–534. https://doi.org/10.1038/nclimate2621.

Metropolis, N., A.W. Rosenbluth, M.N. Rosenbluth, A.H. Teller, and E. Teller. 1953. Equation of state calculations by fast computing machines. *The Journal of Chemical Physics* 21: 1087–1092. https://doi.org/10.1063/1.1699114.

Milne, A.E., M.J. Glendining, R.M. Lark, S.A.M. Perryman, T. Gordon, and A.P. Whitmore. 2015. Communicating the uncertainty in estimated greenhouse gas emissions from agriculture. *Journal of Environmental Management* 160: 139–153. https://doi.org/10.1016/j.jenvman.2015.05.034.

Minunno, F., M. van Oijen, D. Cameron, S. Cerasoli, J. Pereira, and M. Tomé. 2013. Using a Bayesian framework and global sensitivity analysis to identify strengths and weaknesses of two process-based models differing in representation of autotrophic respiration. *Environmental Modelling & Software* 42: 99–115. https://doi.org/10.1016/j.envsoft.2012.12.010.

Minunno, F., M. Van Oijen, D. Cameron, and J. Pereira. 2013. Selecting parameters for Bayesian calibration of a process-based model: A methodology based on canonical correlation analysis. *SIAM/ASA Journal on Uncertainty Quantification* 1: 370–385. https://doi.org/10.1137/120891344.

Murphy, K.P. 2012. *Machine learning: A probabilistic perspective*. Adaptive computation and machine learning series: MIT Press, Cambridge, MA.

Neal, R.M. 1996. Priors for infinite networks. In Neal, R.M. (Ed.), Bayesian Learning for Neural Networks, Lecture Notes in Statistics. Springer, New York, NY, pp. 29–53. https://doi.org/10.1007/978-1-4612-0745-0_2.

Nott, D.J., Y. Fan, L. Marshall, and S.A. Sisson. 2014. Approximate Bayesian Computation and Bayes' Linear Analysis: Toward High-Dimensional ABC. *Journal of Computational and Graphical Statistics* 23: 65–86. https://doi.org/10.1080/10618600.2012.751874.

Ogle, K. 2009. Hierarchical Bayesian statistics: Merging experimental and modeling approaches in ecology. *Ecological Applications* 19: 577–581. https://doi.org/10.1890/08-0560.1.

Ogle, K., and J.J. Barber. 2008. Bayesian DataModel integration in plant physiological and ecosystem ecology. In *Progress in Botany, Progress in Botany*, ed. U. Lüttge, W Beyschlag, and J. Murata, 281–311. Berlin Heidelberg: Springer.

Ogle, K., S. Pathikonda, K. Sartor, J.W. Lichstein, J.L.D. Osnas, and S.W. Pacala. 2014. A model-based meta-analysis for estimating species-specific wood density and identifying potential sources of variation. *Journal of Ecology* 102: 194–208. https://doi.org/10.1111/1365-2745.12178.

O'Hagan, A. 2012. Probabilistic uncertainty specification: Overview, elaboration techniques and their application to a mechanistic model of carbon flux. *Environmental Modelling & Software* 36: 35–48. https://doi.org/10.1016/j.envsoft.2011.03.003.

O'Hagan, A. 2006. Bayesian analysis of computer code outputs: A tutorial. *Reliability Engineering & System Safety* 91: 1290–1300. https://doi.org/10.1016/j.ress.2005.11.025.

O'Neil, C. 2016. *Weapons of math destruction: How big data increases inequality and threatens democracy*, 01 edition. ed. Penguin.

Opitz, T. 2017. Latent Gaussian modeling and INLA: A review with focus on space-time applications. *Journal de la Société Française de Statistique* 158: 62–85.

Patenaude, G., R. Milne, M. Van Oijen, C.S. Rowland, and R.A. Hill. 2008. Integrating remote sensing datasets into ecological modelling: A Bayesian approach. *International Journal of Remote Sensing* 29: 1295–1315. https://doi.org/10.1080/01431160701736414.

Rasmussen, C.E., and C.K. Williams. 2006. *Gaussian processes for machine learning*.

Reichstein, M., G. Camps-Valls, B. Stevens, M. Jung, J. Denzler, N. Carvalhais, and Prabhat, 2019. Deep learning and process understanding for data-driven Earth system science. *Nature* 566: 195. https://doi.org/10.1038/s41586-019-0912-1.

Reyer, C. 2015. Forest productivity under environmental Changea review of stand-scale modeling studies. *Current Forestry Reports* 1: 53–68. https://doi.org/10.1007/s40725-015-0009-5.

Reyer, C.P.O., M. Flechsig, P. Lasch-Born, and M. van Oijen. 2016. Integrating parameter uncertainty of a process-based model in assessments of climate change effects on forest productivity. *Climatic Change* 137: 395–409. https://doi.org/10.1007/s10584-016-1694-1.

Robert, C.P., and G. Casella. 2010. *Monte carlo statistical methods*. New York, NY: Springer.

Robert, C.P., N. Chopin, and J. Rousseau. 2009. Harold Jeffreys's theory of probability revisited. *Statistical Science* 24: 141–172. https://doi.org/10.1214/09-STS284.

Rollinson, C.R., Y. Liu, A. Raiho, D.J.P. Moore, J. McLachlan, D.A. Bishop, A. Dye, J.H. Matthes, A. Hessl, T. Hickler, N. Pederson, B. Poulter, T. Quaife, K. Schaefer, J. Steinkamp, and M.C. Dietze. 2017. Emergent climate and CO2 sensitivities of net primary productivity in ecosystem models do not agree with empirical data in temperate forests of eastern North America. *Global Change Biology* 23: 2755–2767. https://doi.org/10.1111/gcb.13626.

Rougier, J. 2008. Efficient emulators for multivariate deterministic functions. *Journal of Computational and Graphical Statistics* 17: 827–843. https://doi.org/10.1198/106186008X384032.

Rougier, J. 2007. Probabilistic inference for future climate using an ensemble of climate model evaluations. *Climatic Change* 81: 247–264. https://doi.org/10.1007/s10584-006-9156-9.

Roustant, O., D. Ginsbourger, and Y. Deville. 2012. DiceKriging, DiceOptim: Two R packages for the analysis of computer experiments by kriging-based metamodeling and optimization.

Schlesinger, W.H., M.C. Dietze, R.B. Jackson, R.P. Phillips, C.C. Rhoades, L.E. Rustad, and J.M. Vose. 2016. Forest biogeochemistry in response to drought. *Global Change Biology* 22: 2318–2328. https://doi.org/10.1111/gcb.13105.

Schneiderbauer, S., and D. Ehrlich. 2004. *Risk, hazard and people's vulnerability to natural hazards* 42.

Shachter, R.D., and C.R. Kenley. 1989. Gaussian influence diagrams. *Management Science* 35: 527–550. https://doi.org/10.1287/mnsc.35.5.527.

Simpson, A.H., S.J. Richardson, and D.C. Laughlin. 2016. SoilClimate interactions explain variation in foliar, stem, root and reproductive traits across temperate forests. *Global Ecology and Biogeography* 25: 964–978. https://doi.org/10.1111/geb.12457.

Sivia, D., and J. Skilling. 2006. *Data a: A Bayesian tutorial*, 2 edition. ed. Oxford University Press, U.S.A., Oxford.

Smith, R., J. Dick, H. Trench, and M. Van Oijen. 2012. Extending a Bayesian Belief Network for ecosystem evaluation. In: Conference Paper of the 2012 Berlin Conference of the Human Dimensions of Global Environmental Change on "Evidence for Sustainable Development", 5–6 October 2012, Berlin, Germany.

Soyer, R. 2018. Kalman filtering and sequential Bayesian analysis. *Wiley Interdisciplinary Reviews: Computational Statistics* 10: e1438. https://doi.org/10.1002/wics.1438.

Spence, M.A., J.L. Blanchard, A.G. Rossberg, M.R. Heath, J.J. Heymans, S. Mackinson, N. Serpetti, D.C. Speirs, R.B. Thorpe, and P.G. Blackwell. 2018. A general framework for combining ecosystem models. *Fish and Fisheries* 19: 1031–1042. https://doi.org/10.1111/faf.12310.

Spiegelhalter, D., M. Pearson, and I. Short. 2011. Visualizing uncertainty about the future. *Science* 333: 1393–1400. https://doi.org/10.1126/science.1191181.

Sutton, M.A., D. Simpson, P.E. Levy, R.I. Smith, S. Reis, M. van Oijen, and W. de Vries. 2008. Uncertainties in the relationship between atmospheric nitrogen deposition and forest carbon sequestration. *Global Change Biology* 14: 2057–2063. https://doi.org/10.1111/j.1365-2486.2008.01636.x.

ter Braak, C.J.F., and J.A. Vrugt. 2008. Differential evolution Markov chain with snooker updater and fewer chains. *Statistics and Computing* 18: 435–446. https://doi.org/10.1007/s11222-008-9104-9.

Tokmakian, R., P. Challenor, and Y. Andrianakis. 2012. On the use of emulators with extreme and highly nonlinear geophysical simulators. *Journal of Atmospheric and Oceanic Technology* 29: 1704–1715. https://doi.org/10.1175/JTECH-D-11-00110.1.

Van Oijen, M. 2020. CAF2014. https://doi.org/10.5281/zenodo.3608877.

Van Oijen, M. 2017. Bayesian methods for quantifying and reducing uncertainty and error in forest models. *Current Forestry Reports* 3: 269–280. https://doi.org/10.1007/s40725-017-0069-9.

Van Oijen, M., G.I. Ågren, O. Chertov, S. Kellomäki, A. Komarov, D. Mobbs, and M. Murray. 2008. 4.4 Evaluation of past and future changes in European forest growth by means of four process-based models. *Causes and Consequences of Forest Growth Trends in Europe: Results of the Recognition Project* 21: 183–199.

Van Oijen, M., J. Balkovic, C. Beer, D.R. Cameron, P. Ciais, W. Cramer, T. Kato, M. Kuhnert, R. Martin, R. Myneni, A. Rammig, S. Rolinski, J.-F. Soussana, K. Thonicke, M. Van der Velde, and L. Xu. 2014. Impact of droughts on the carbon cycle in European vegetation: A probabilistic risk analysis using six vegetation models. *Biogeosciences* 11: 6357–6375. https://doi.org/10.5194/bg-11-6357-2014.

Van Oijen, M., C. Beer, W. Cramer, A. Rammig, M. Reichstein, S. Rolinski, and J.-F. Soussana. 2013. A novel probabilistic risk analysis to determine the vulnerability of ecosystems to extreme climatic events. *Environmental Research Letters* 8: 015032. https://doi.org/10.1088/1748-9326/8/1/015032.

Van Oijen, M., and D. Cameron. 2020. BASFOR. https://doi.org/10.5281/zenodo.3608882.

Van Oijen, M., D. Cameron, K. Butterbach-Bahl, N. Farahbakhshazad, P.-E. Jansson, R. Kiese, K.-H. Rahn, C. Werner, and J. Yeluripati. 2011. A Bayesian framework for model calibration, comparison and analysis: Application to four models for the biogeochemistry of a Norway spruce forest. *Agricultural and Forest Meteorology* 151: 1609–1621. https://doi.org/10.1016/j.agrformet.2011.06.017.

Van Oijen, M., D. Cameron, P.E. Levy, and R. Preston. 2017. Correcting errors from spatial upscaling of nonlinear greenhouse gas flux models. *Environmental Modelling & Software* 94: 157–165. https://doi.org/10.1016/j.envsoft.2017.03.023.

Van Oijen, M., M.G.R. Cannell, and P.E. Levy. 2004. Modelling biogeochemical cycles in forests: State of the art and perspectives. Towards the sustainable use of European forests-Forest ecosystem and landscape research: scientific challenges and opportunities, edited by: Andersson, F., Birot, Y., and Päivinen, R., European Forest Institute, Joensuu, Finland 157–169.

Van Oijen, M., M. Höglind, D. Cameron, and S. Thorsen. 2015. BASGRA_2014. https://doi.org/10.5281/zenodo.27867.

Van Oijen, M., C. Reyer, F. Bohn, D. Cameron, G. Deckmyn, M. Flechsig, S. Härkänen, F. Hartig, A. Huth, A. Kiviste, P. Lasch, A. Mäkelä, T. Mette, F. Minunno, and W. Rammer. 2013. Bayesian calibration, comparison and averaging of six forest models, using data from Scots pine stands across Europe. *Forest Ecology and Management* 289: 255–268. https://doi.org/10.1016/j.foreco.2012.09.043.

Van Oijen, M., J. Rougier, and R. Smith. 2005. Bayesian calibration of process-based forest models: Bridging the gap between models and data. *Tree Physiology* 25: 915–927. https://doi.org/10.1093/treephys/25.7.915.

Van Oijen, M., and A. Thomson. 2010. Toward Bayesian uncertainty quantification for forestry models used in the United Kingdom Greenhouse Gas Inventory for land use, land use change, and forestry. *Climatic Change* 103: 55–67. https://doi.org/10.1007/s10584-010-9917-3.

Van Oijen, M., and M.A. Zavala. 2019. Probabilistic drought risk analysis for even-aged forests. In *Climate Extremes and Their Implications for Impact and Risk Assessment*, ed. J. Sillmann, S. Sippel, and S. Russo, 159–176. Elsevier.

Welsh, A., A. Peterson, and S. Altmann. 1988. The fallacy of averages. *The American Naturalist* 132: 277–288.

Wikle, C.K., and J.A. Royle. 1999. Space-time dynamic design of environmental monitoring networks. *Journal of Agricultural, Biological, and Environmental Statistics* 4: 489–507.

Williams, C.K., and C.E. Rasmussen. 1996. Gaussian processes for regression. In: NIPS'95: Proceedings of the 8th International Conference on Neural Information Processing Systems. pp. 514–520.

Williams, D.R., P. Rast, and P.-C. Bürkner. 2018. Bayesian meta-analysis with weakly informative prior distributions. https://doi.org/10.31234/osf.io/7tbrm.

Williams, P.J., and M.B. Hooten. 2016. Combining statistical inference and decisions in ecology. *Ecological Applications* 26: 1930–1942.

Wood, S. 2006. *Generalized additive models: An introduction with R*. CRC press.

Xu, Y., J. Choi, S. Dass, and T. Maiti. 2016. *Bayesian prediction and adaptive sampling algorithms for mobile sensor networks: Online environmental field reconstruction in space and time, SpringerBriefs in control*. Automation and Robotics: Springer International Publishing.

Yeluripati, J.B., M. van Oijen, M. Wattenbach, A. Neftel, A. Ammann, W. Parton, and P. Smith. 2009. Bayesian calibration as a tool for initialising the carbon pools of dynamic soil models. *Soil Biology and Biochemistry* 41: 2579–2583. https://doi.org/10.1016/j.soilbio.2009.08.021.

Young, P. 1998. Data-based mechanistic modelling of environmental, ecological, economic and engineering systems. *Environmental Modelling & Software* 13: 105–122. https://doi.org/10.1016/S1364-8152(98)00011-5.

Index

A

Abbreviations, 177
Accept-Reject (A-R) algorithm, 31, 33, 58, 70, 72
Activation function, 144, 145
Adaptive algorithms for MCMC, 43, 70
Adaptive sampling, 170, 172
Akaike Information Criterion (AIC), 136
Approximate Bayesian Computation (ABC), 136
Approximation to Bayes, 135
Autoregressive model (AR)
 – autoregressive moving average model (ARMA), 153
 – conditional autoregressive model (CAR), 161
 – simultaneous autoregressive model (SAR), 161

B

Backpropagation algorithm, 148
Basis functions, 139
Bayes factor, 82
Bayesian Belief Network (BBN), 107
Bayesian Calibration (BC), 11, 16, 39–42, 44, 46, 49, 63, 65–67, 69–75, 77
Bayesian Decision Theory (BDT), 120, 129, 131
Bayesian expected loss, 131
Bayesian expected utility, 131
Bayesian Hierarchical Modelling (BHM), 80, 121–123, 128, 162, 174
Bayesian Model Averaging (BMA), 83
Bayesian Model Comparison (BMC), 82
Bayesian Network (BN), 107, 120
Bayes' theorem, 1, 3–6

B

Bernoulli distribution, 34, 35
Borrowing strength, 127
Burn-in of MCMC, 44–46, 74

C

Causality, 108, 120
Classification, 142
Clustering, 142
Coherent probability assigment, 17, 132
Computational demand, 174
Concave functions, 166, 167
Conditional independence, 108, 114, 120, 161
Conditional multivariate Gaussian, 49, 55
Conditional probability table, 120
Conjugate distributions, 29, 159
Convergence of MCMC, 33, 35, 46, 69, 74
Convex functions, 166, 167
Correlation coefficient, 50
Correlation length, 57, 96, 100, 104, 105, 118
Cost function, *see also* Likelihood function
Covariance function, 94–101, 104, 118, 151–154, 161, 162, 169, 171
 – separable covariance function, 169
Covariance matrix, 25, 26, 37, 41, 52, 53, 55–57, 64, 96–98, 100, 107, 109, 111–118, 120, 161, 164, 167, 183
Cox postulates, v

D

Data assimilation, 154–160
Decision theory, *see* Bayesian decision theory
Decision tree, 142

© Springer Nature Switzerland AG 2020
M. van Oijen, *Bayesian Compendium*,
https://doi.org/10.1007/978-3-030-55897-0

Deep learning, 143, 149
Density estimation, 75
Design matrix, 52, 55, 96
Dimensional analysis, 179
Dimensionality reduction, 142, 143
Directed Acyclic Graph (DAG), 107, 108,
    133
Discrepancy (model structural error), 70, 74,
    77, 81, 90, 91, 174
Drivers, *see* Model drivers
Dynamical system, 54

E
Empirical Bayes, 124
Emulation, *see* Emulator
Emulator, 93–98, 100–104, 174
    – statistical emulator, 93, 174
Ensemble modelling, 81, 91
Entropy, 18, 19
Error, 7–10, 12–14, 16
    – error propagation, 14
    – measurement error, 24–26
    – representativeness error, 24
    – spatial upscaling error, 166
    – systematic error, 24, 71, 74
Evidence, 2–5
Expolinear (equation) model, 63–66, 84, 85
Exposure, 131

F
Fallacy of averages, 166
Filtering, 155, 157, 160
Fixed effect, 139

G
Gaussian Bayesian Network (GBN), 107–
    110, 112–120
Gaussian Markov Random Field (GMRF),
    161, 162
Gaussian Network (GN), 49, 107–110, 112–
    120
Gaussian Process (GP), 10, 12, 94, 149, 162,
    169
Gelman-Rubin statistic, 69
Generalized   Additive   Mixed   Model
    (GAMM), 138–140
Generalized Additive Model (GAM), 137–
    140
Generalized Linear Mixed Model (GLMM),
    138–140

Generalized Linear Model (GLM), 137–140,
    144
Geostatistics, 118, 119, 162, 163, 165, 166
Gibbs sampling (MCMC algorithm), 49, 58–
    61, 70, 140
Graphical Model (GM), 57, 80, 107, 114,
    119, 120, 133

H
Hazard, 129–131
Hessian matrix, 167
Hidden Markov Model (HMM), 120, 157
Hierarchical modelling, *see* Bayesian hierar-
    chical modelling
Hyperparameter, 101, 104, 121–128, 139,
    162, 164, 166, 170
Hyperprior, 121, 126

I
Identifiability of parameters, 46
Importance sampling, 32, 70
Information criteria, 136
Integrated likelihood, 81–83, 86, 136
Integrated Nested Laplace Approximation
    (INLA), 136
Interpretation, 77, 141

J
Jacobian matrix, 160
JAGS, 49, 60, 61, 125, 126
Jeffreys prior, 20

K
Kalman Filtering (KF), 49, 62, 155, 157,
    170, 172
    – ensemble Kalman filter (EnKF), 160
    – extended Kalman filter (EKF), 160
Kalman gain, 55, 155, 157, 159
Kernel function, *see* Covariance function
Kriging, 118, 119

L
Law of total probability, 181
Least squares estimation, 49
    – generalised least squares (GLS), 49, 52
    – ordinary least squares (OLS), 49, 50
    – weighted least squares (WLS), 49, 53
Likelihood function, 3, 4, 6, 16, 23–27, 70,
    71, 74

Linear algebra, 180
Linear Bayes (LB), 136
Linear (regression) modelling, 57, 137–140
  – linear mixed model (LMM), 139
Link function, 138, 139
Logistic regression, 138, 144
Log-transformation (need for), 40

M

Machine learning, 141, 142, 147–149, 173–
  175
Marginal distribution, 46, 59, 77, 98, 108,
  127, 136, 146, 165
Markov Chain Monte Carlo sampling
  (MCMC), 32–37, 39–47, 64, 66, 67,
  135, 174
Markov Random Field (MRF), 107
Mathematics, 179
Maximum A Posteriori parameter estimate
  (MAP), 75
Maximum Entropy (MaxEnt), 18, 19
Maximum Likelihood Estimation (MLE),
  17, 124, 136, 139, 140
Mean Squared Deviation (MSD), 78
Measurement equation, 7
Meta-analysis, 122
Metropolis algorithm for MCMC, 35, 36, 40,
  41, 58, 145
Mixed modelling, *see* Bayesian hierarchical
  modelling
Mixing of MCMC, 35
Model, 7–18
  – deterministic model, 10
  – dynamic model, 9, 11, 39, 63
  – empirical model, 10, 11
  – process-based model (PBM), 9, 11, 63
  – statistical model, 10, 11
  – stochastic model, 10
Model drivers, 12, 15
Modelling Equation, 7, 8, 12, 16
Model structural error, *see* Discrepancy
Moments, 19
Monte Carlo methods, 14
Moving average model (MA), 153
Multilevel modelling, *see* Bayesian hierar-
  chical modelling
Multimodal distribution, 37, 84

N

Neighborhood, 161
Neural network models, 141
  – convolutional network, 144

  – feedforward neural network, 143–145,
  149
  – recurrent neural network, 144
Normal Equations, 50, 51
Notation, 177
Nugget, 104, 162–165

O

Observational operator matrix, 54, 157, 160
Occam's razor, 82
Overfitting, 147

P

Parameter, 7–16
  – parameter screening, 34, 72
Particle filtering, 160
Penalized regression, 140, 147
Penalty function, 140
Point-support, 166
Posterior probability distribution, 3, 29, 39,
  40, 45, 46
Precision matrix, 109, 111, 116, 161
Predictive distribution, 46
Principal Component Analysis (PCA), 143
Prior probability distribution, 2, 17, 19, 73,
  75
  – Jeffreys prior, 20
  – hierarchical prior, 20
Probabilistic network, 107
Probabilistic risk analysis, 130, 175
Probability, 8–10, 12–16
  – conditional probability, 3
Probability distribution, 181
  – beta, 13, 72, 131
  – exponential family, 138
  – Gaussian, 9, 10, 17, 19, 25, 29, 36, 54,
  107, 151, 161
  – inverse Gamma, 165
  – scaled inverse χ^2, 165
  – t, 166
Probability theory, v, vii, 1, 2, 181
Process, 8–15
  – Bernoulli stochastic process, 35
  – Gaussian process, *see* Gaussian process
  (GP)
  – stochastic process, 10, 12, 57
Product rule of probability, 181
Proposal distribution for MCMC, 34, 35, 42–
  45, 47, 70, 73

R
R, 185, 186
– R-functions defined in this book
     * EXPOL5, 64, 65
     * EXPOL6, 84
     * EXPOL6s, 100
     * fNN.11, 149
     * GaussCond, 55, 56, 113
     * GaussMult, 58
     * GP.est, 97
     * GP.pred, 97, 98
     * GP.pred.0, 152
     * KalmanGain, 55, 171
     * KalmanGainSVD, 159
     * logLik, 41
     * logLikList, 65
     * logPost, 41
     * logPostList, 65
     * logPrior, 41
     * Metropolis, 35
     * MetropolisLogPost, 41
     * precMatrix, 110
     * VcondR, 110
– R-packages
     * BACCO, 104
     * caret, 149
     * DiceKriging, 104
     * FD, 18
     * gamm4, 140
     * geoR, 96, 98, 163
     * lme4, 140
     * mgcv, 140
     * rjags, 60, 125
Random, 8, 10, 24
– random effect, 139
– random function, 95
– random variable, 8, 95
Random forest, 142, 148
Regression, 50, 51, 53–59, 61, 62, 138–140,
     142–144, 149
Regression coefficient, 109–114, 116, 117
Regularization, 147
Rejection sampling, *see* Accept-Reject algo-
     rithm
Reporting, 79, 80
Residual, 8, 9
Reversible Jump MCMC (RJMCMC), 83
Risk, 129–131, 133
Risk analysis, *see* Probabilistic risk analysis

S
Scaling parameters, 20
Science Equation, 8
Sequential Bayesian updating, 71, 182
Shrinkage to the mean, 127
Singular Value Decomposition (SVD), 159
Smooth function, 139
Software (Bayesian), 187
Spatial modelling, 161, 162, 166
Spatio-temporal modelling, 169
Spin-up of model, 73
Splines, 139
State estimation, 54, 154–160
State-space model, 154, 155, 160
Stationarity, 153
Structural equation modelling, 120
Supervised learning, 142
Support Vector Machine (SVM), 142
Systems identification, 104, 174

T
Taylor expansion, 12, 166
Terminology, 8, 10
Time series model, 151, 153, 155, 157, 159
Trace plot, 34, 35, 42–45

U
Uncertainty, 7–16
– structural uncertainty, 89, 90, 92
Unsupervised learning, 142, 143
Utility, *see* Bayesian expected utility

V
Value of information, 132, 133
– expexted value of partial information,
     132
– expexted value of perfect information,
     132
Variability, 8, 9
Variance, 18–20
– average predictive variance (APV),
     170, 172
– conditional variance, 109–114, 117
– unconditional variance, 56, 109, 111
Varying intercept, varying-slope model, 128
Visualization, 77, 80
Vulnerability, 129–131

Printed in the United States
by Baker & Taylor Publisher Services